Tolib Turaev

An effective method for stabilizing gas condensate

Tolib Turaev

An effective method for stabilizing gas condensate

Monograph

ScienciaScripts

Imprint

Cover image: www.ingimage.com

This book is a translation from the original published under ISBN 978-620-8-41700-0.

Publisher:
Sciencia Scripts
is a trademark of
Dodo Books Indian Ocean Ltd. and OmniScriptum S.R.L publishing group

120 High Road, East Finchley, London, N2 9ED, United Kingdom
Str. Armeneasca 28/1, office 1, Chisinau MD-2012, Republic of Moldova, Europe
Managing Directors: Ieva Konstantinova, Victoria Ursu
info@omniscriptum.com

Printed at: see last page
ISBN: 978-620-8-59619-4

MINISTRY OF HIGHER EDUCATION OF THE REPUBLIC OF UZBEKISTAN
EDUCATION, SCIENCE AND INNOVATION

TOLIB BOZOROVICH TURAEV

EFFICIENT METHOD OF GAS CONDENSATE STABILIZATION

On the rights of the manuscript
UDC: 665.62

The monograph deals with the development of theoretical foundations study of technological schemes of gas and gas condensate stabilization processes, operating modes of individual apparatuses and their design features, the characteristics of different types of raw materials give the main directions of obtaining different products and, accordingly, will show the main directions of development of modern gas chemical and oil and gas processing industry. This is especially important in the issues of modernization of enterprises of this industry, as well as in the issues of obtaining methane homologues.

Gas condensate is an important product released during gas production. Stable condensate is used to produce motor fuels and valuable chemical products. So far, the schemes of its separation at most gas processing plants are far from perfect, so the issues of increasing the productivity of industrial plants, their improvement, improvement of their technology and environmental safety are very relevant.

The process of stabilization of gas condensate are an important step for obtaining broad fractions of hydrocarbon fuel, stable gas condensate is light, processed into a finished product, and on transporting it to refineries.

TABLE OF CONTENTS

INTRODUCTION

With the growing requirements for motor vehicles (more powerful engine, payload, comfort and less fuel consumption, dimensions, etc.), the requirements for the quality of motor fuel are increasing [1]. In recent years, the tendency of fuel manufacturers to create their compositions allowing to provide high detonation resistance, calorific value, more complete combustion with less soot formation, etc. is increasing. [2]. To ensure the above requirements, the closest way is to obtain fuels containing low-molecular alcohols and esters. The most known additives used instead of tetraethyl lead detonators in fuels are iso esters (methyl tert-butyl, ethyl tert-butyl and tert-butylisoamyl esters). However, their production is associated with certain difficulties; transportation, storage and certain negative properties make them vulnerable for their use in fuels. Therefore, in order to obtain low-molecular oxygen-containing hydrocarbons we resort to oxidation of light fractions of gas condensate hydrocarbons on catalysts MnO2 or H3BO3 [3] steam-condensate contact of hydrocarbons in the environment of depleted air oxygen.

The study of technological schemes of gas and gas condensate stabilization processes, operating modes of individual apparatuses and their design features, characterization of different types of raw materials give the main directions of obtaining different products and, accordingly, will show the main directions of development of modern gas chemical and oil and gas processing industry. This is especially important in the issues of modernization of enterprises of this industry, as well as in the issues of obtaining methane homologues.

Gas condensate is an important product released during gas production. Stable condensate is used to produce motor fuels and valuable chemical products. So far, the schemes of its separation at most gas processing plants are far from perfect, so the issues of increasing the productivity of industrial plants, their improvement, improvement of their technology and environmental safety are very relevant.

The process of stabilization of gas condensate is an important stage for obtaining broad fractions of hydrocarbon fuel, stable gas condensate is light, processed into a finished product, and on its transportation to oil refineries such as Bukhara refinery and Fergana refinery. Proceeding from the above, stabilization of gas condensate is considered an important stage in the field.

In this connection, the present master's thesis, devoted to the characteristics of the gas condensate stabilization unit, operating at Mubarek GPP modernization of the existing unit, improving the quality of output products and reducing the cost of energy consumption, is relevant.

The main purpose of this monograph, improving the efficiency of stabilization of gas condensate properties and the development of improved technological schemes of gas condensate stabilization plants and process intensification.

In this regard, the following objectives were set and solved:

- to make a scientific and practical analysis of the known literature materials devoted to research and development of practical experience of stabilization of gas condensate (GC), as well as the application of GC;
- theoretical issues of chemistry and rheology of HA are considered;
- To develop methods of analysis and object of research -GC, as well as auxiliary materials, extractants and experimental means;
- To develop an improved technological scheme of gas condensate stabilization unit and process intensification;
- Improvements in the efficiency of stabilization of gas condensate properties.

Development of an effective method of stabilization of unstable gas condensate. Improvement of the current process of gas condensate stabilization.

The first chapter shows the technology of stabilization of gas condensate by recirculation of gas degassing, stabilization of condensate multistage degassing, and fractionation of gas condensate at the gas separation plant to obtain solvents and motor fuel. To date, in this direction are conducted at the Institute of General and Inorganic Chemistry in the Laboratory of Gas Condensates under the guidance of DrAlimov A. Alimov.

The main methods used to determine the properties of gas condensate products, Fractional composition in the ARN-2 apparatus, Conditional viscosity, determination of aromatic hydrocarbons, determination of flash point in a closed crucible, sulfur content by burning in a lamp, presence of water-soluble acids and alkalis, determination of acid number. Infrared radiation, liquid and gas chromatography and elemental analysis are also used to determine the properties of gas condensate products.

On the basis of the conducted researches and obtained results the technological scheme, steadily working in a wide range of changes of fractional composition of initial raw material, was developed and realized during reconstruction.

CHAPTER I. LITERATURE REVIEW. GENERAL CHARACTERISTIC OF GAS CONDENSATE STABILIZATION PROCESS

1. Gas condensates of Uzbekistan and their properties.

During the years of independence of our country, the oil and gas processing industry has made unprecedented steps to develop new gas and oil fields and their processing. With great support of the industry by the Government of the Republic under the leadership of I. A. Karimov, the Bukhara Oil Refinery (BOR), Shurtan Gas and Chemical Complex (SGCC), a number of gas processing enterprises were developed, construction of large gas storage facilities was completed, and some production facilities of oil refineries were reconstructed. As a result, fuel independence of the economy of the Republic of Uzbekistan was achieved [3].

A special success can be noted with satisfaction with the launch and development of the SHGC for the first time, the production of chemical products of polyethylene (all their varieties) from natural gas - ethane - has been accelerated in the region. In general, 4.5 - 5.0 bln. m^3/year of raw gas will be processed, which will be used to produce [3]:

1. Marketable gas (methane - 99%) — 3.5-4.0 billion m^3/year
2. Gas condensate (stable) — 90-110 thousand tons/year;
3. Ethane (process gas) — 160-165 thousand tons per year
4. Ethylene (to polyethylene) — 140-145 thousand tons/year
5. Liquefied gas (propane-butane) — 110-135 thousand tons per year

6. Gas sulfur (99.9% purity) 4.5-5.5 thousand tons/year

However, it should be noted that most of the production (85% of marketable gas, gas condensate, liquefied gas, etc.) of the complex still works for fuel and from a small share of raw materials - ethane is obtained ethylene, from which polyethylene is synthesized - 125 thousand tons/year. Hoping for the future of SHGCC, as the President of the Republic noted at the presentation, "... it is necessary to obtain a considerable list of chemical materials from the obtained gas...," [4], i.e., developing the above thought: the chemical complex has a wide field of action in the development of chemical processing of all components of natural gas into chemical intermediates and materials.

Tremendous possibilities of SHGCC as the of the only gas chemical production in the republic, embodying the latest technologies, gas purification, gas separation, gas converter (pyrolysis furnace), cold production, gas storage and other auxiliary units, promote self-development - mastering new technologies, production of ethylene oxide, ethanolamines, ethylene glycols, propylene etc.

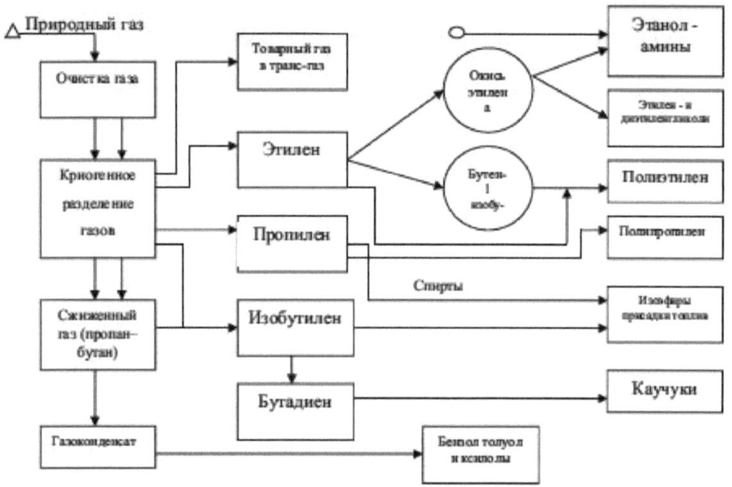

Fig.1 Current and prospective scheme of SHCP development for the production of gas chemical synthesis products.

Δ-acting technologies and products; -prospe○ive technologies and products at UZ-KOR-GAS-Chemical (Surgilinsk MCC)

Polyethylene production gives import substitution of more than 300 items of chemical products manufactured in our country, and further stages of gas chemical synthesis product will provide progressive forces for the development of chemical materials science [5]. Early scientific research laid the foundation for new practical developments on technologies for chemical intermediates - benzene, toluene and xylene (BTK) [6], compound gasoline [7], ashless additives [8,9], environmentally friendly oil-extraction solvents [10], varieties of solvents for mechanical engineering [11,12], etc. [13,14].

At the same time scientific-theoretical substantiation and technical documentation for some of the listed products of chemical utilization of gas condensate have been found and created. For chemical processing of gas and condensates the composition and properties of the exploited fields of the republic have been studied (Table 1).

Table 1.

Composition and properties of gas and refinery gases

Gas	Air specific gravity, g/m^3	Hydrocarbon composition, %				Amount and other composition, %
		CH_4	C_2H_6	C_3H_8	C_4H_{10}	
Raw natural	640	76,4	5,1	3,2	1,7	13-17
Commodity natural	538	99,3	0,3	0,1	-	-
Technological	660	0,2	58,5	30,2	11,3	0,5-1,0
Liquefied	650	0,1	0,4	60,7	35,3	2,5-3,0

As can be seen, the average composition of natural gases is rich in process gases, and they contain mainly ethane and liquefied gas fractions. These components of natural gas, as it is known from long-term experience, are the cleanest and technologically favorable semi-products for their deep chemical processing, which is confirmed by production practice in GPPs. This is a vivid example of gas transformation into chemical material and further - the key to the development of chemical materials science.

As a result of long-term study of gas condensates from exploited natural gas fields, it can be concluded that they meet the following chemical processing requirements:

- low-sulfur (bound) compared to petroleum;
- highly aromatic (12-27 May %);
- relatively light and mobile hydrocarbon mixtures (35-380°C);
- light-colored and easily recyclable.

These totals are sufficient conditions for chemical utilization and processing into intermediates.

Table 2.

Indicators of properties of gas condensates of Uzbekistan

Fishing	HC production, thousand tons for 2001.	Temperature, °C		Physical properties		Group composition hydrocarbons, *wt* %			Bound sulfur content, %	GC color by nodomet-rica scale, mg J/L
		N. K.	C.c..	d_4^{20} kg/m^3	Π_D^{20}	Paraffin	Naphthenic	Aromatic		
	2000	45	385	764	1,4592	62,1	29,1	8,8	0,2-0,3	28
Shurtan	800	35	360	740	1,4358	47,0	26,0	27,0	,08-0,1	12
Mubarek s/n	380	40	350	738	1,4385	56,2	29,3	14,5	,15-0,2	19
Bukhara group of meso-births	410	50	380	787	1,4486	39,1	36,7	24,2	,25-0,4	30

Summa and other deposits	650	55	390	810	1,467 5	42, 5	40,8	16,7	0,3-0,5	35

According to the research data (Table 2), it can be judged that by the volume of production and group composition the given gas condensates are unique and quite suitable for use in petrochemistry. The set of indicators was analyzed for stable gas condensates shipped for processing into fuel, their compliance with the requirements to them of technical specifications was checked.

During low-temperature separation and stabilization of gas condensate, light hydrocarbons are carried away, which are extracted as a cube residue during propane-butane separation - broad fraction of light hydrocarbons (NGL). NGL are characterized by low boiling point with saturated vapor pressure of 512.8 *MPa*, absence of water and mechanical impurities, very low (traces) content of sulfur and aromatic hydrocarbons, high mobility, transparency, etc.

Table 3.

Qualitative and quantitative composition of solvents for extraction of vegetable oils.

Components	Boiling point, °C	Solvents		
		GK-30/115 NGL	GC-ER 30/85	GC-ER 65/85
H-pentane	36.1	0.35	0.75	0.92
2.3-Dimethylbutane	58.0	2.46	2.29	2.08
2-Methylethane	60.0	3.20	14.47	19.60
H-hexane	68.3	36.50	37.24	29.97
Methylcyclopentane	71.8	10.90	11.78	9.24
2.4-Dimethylpentane	80.5	2.95	2.90	3.12
Benzene	80.1	0.05	0.07	0.08
1-methylhexane	85.2	0.92	1.20	1.01
3-Methylhexane	84.0	0.31	0.38	0.41
1.2-Dimethylcyclopropane	78.2	0.82	0.90	0.94
Methylcyclobutane	76.4	0.60	0.55	0.60

*Other hydrocarbons	30-115	4.54	3.25	0.10
C_2-C^1_{10}	<<	3.25	-	-
*-light NGL constituents, which are included in GC-30/115 and GC-ER 30/85.				

Taking into account these properties, the technology of obtaining gas condensate extractant-solvent (GC-ER 65/85) was developed, i.e. the parameters of the process of extraction of hydrocarbon fractions of 65-85°C from NGL were established. To characterize the solvent extractant of vegetable oils, a complete analysis of the composition of hydrocarbon fractions was carried out (Table 3).

The absence of foreign aggressive components in the composition of solvents was noted, and if even traces of sulfur and aromatic hydrocarbons were found, they are many times lower than the requirements of known analogs. The obtained solvents were tested at extraction of cotton oil from cake, which contributed to acceleration of the extraction process in 1.5 times at qualitative selection of oils with provision of its high ecological purity.

The test results of the developed solvent provide adhesion of pre-degreased samples under primers like LK-070 and EP 0215 up to 95%, enamels PF-115 and PF-223 96%, sealants UZOMES 5 under layer P-9 up to 98% and detectability of defects at capillary inspection of parts 100%.

Table 4.

Indicators of gas condensate solvent (GK-RS 80/120)

Normative indicators and units.	Norms
Density at 20 °C, g/cm^3	0,73
Fractional composition:	
Boiling point, °C	80
93.0 % must be distilled at a temperature of at least	110
98.0 % must be distilled at a temperature of at least	120
Residue in the flask after distillation 6%, not more	1,5
Bromine number g.bromine/100 cm^3, max.	0,09

Oil slick test (30 min.)	Withstands
Content of water-soluble acids and alkalis	Absent
Mechanical impurities and water content	Below normal
Tetraethyl lead content	Below normal
Flash point, °C, not lower than	-
Autoignition temperature, °C	270
Electrical conductivity of the shift peak/m, not less than	50

NGL is also a valuable raw material for obtaining solvent used in aircraft construction, it is used in defectoscopy of engine parts, for preparation of adhesives, sealants, etc., and it has special requirements, which were achieved during bench-scale development of the process of obtaining solvent GC-RS 80/120°C (Table 4).

High content (32 - 34 wt.%) of (BTK) in the target fraction (65 - 125 ^{0}C) of Mubarek gas condensate allowed to develop their extraction on selective extractant. BTK, which is of interest as a component of motor fuel, is a semi-product for the synthesis of a number of chemical products and materials. Chemically pure benzene and toluene were obtained from BTK with corresponding indicators on their quality certificate.

Raw material capabilities of gas condensate and semi-products of operating plants allow to obtain high-quality grades of compound gasolines. At the same time straight-run and reforming gasoline from gas condensate filled with newly synthesized ashless additives (ethyltretbutyl, diisopropyl and ditretbutyl ethers) made it possible to improve fuel performance.

Table

Comparative performance of petroleum and gas condensate solvents.

Indicator	Known petroleum solvents	Hydrocarbon solvents from condensate gas

	White spirit	Sol-Vent	Nefras S-3 70/98	Nefras C-4 130/210	GK-LKM 120/220	GK-LKM 120/160	GK-RE 65/85
1	2	3	4	5	6	7	8
Density at 20°C, kg/m^3	795	865	695	790	790	845	680
Boiling point,^{0}C	130	120	72	130	120	120	63
End boiling point,^{0}C	210	160	98	98	220	160	98/85
Residue in the flask,%	2	8	1	2	0,5	-	2

Continuation of Table 5

1	2	3	4	5	6	7	8
Mass fraction of aromatics,%	16	65	4	16	18	42	-
Mass fraction of sulfur, %	0,02	0,04	0,025	0,08	0,01	-	0,01
Flash point	30	20	12	32	30	24	13
Xylene volatility	4,5	5-6	2,5-3	6	4-6	3-4	2-2,6
Solvent color	Light red	Light yellow	Proz-rachny	Light yellow	Proz-rachny	Proz-rachny	Proz-rach

Significant progress has been made in the development of gas condensate solvents for paint and varnish materials [15], which are better than known solvents derived from oil (Table 5)

Thus, summarizing the results of the research, the following conclusions can be drawn:

- comparative and qualitative composition of natural gases (methane, ethane, propane, butane and gas condensates) helps to obtain deeper and more efficient technological developments that need to be developed to production realization;

- technologies for production of a number of compounds (benzene, toluene and BTK mixture), compound gasolines and gas-condensate solvents were developed, which should be mastered at UDP Shurtanneftegaz, Mubarekneftegaz and ShGHK;

- unlimited demand both in the republic and abroad for the reduced components of gases and gas condensates, the realization of which makes it possible to achieve socio-economic efficiency [3,4].

2. Processes of stabilization of gas condensate properties

To obtain stable condensate, rectification and multi-stage degassing (separation) processes are mainly used, either separately or in combination.

Stabilization of condensate by multistage degassing is based on decreasing solubility of light components in C_{5+} hydrocarbons increasing temperature and decreasing pressure. Different solubility of components ensures their selective separation from the liquid phase.

One-, *two-* and three-stage degassing (separation) schemes can be used for condensate stabilization.

Below are the results of studies to determine the efficiency of the condensate stabilization process using each of the scheme options. The efficiency criterion is the degree of heavy hydrocarbons (C_5H_{12+}) distribution between separation gases and stable condensate. The compositions of the investigated unstable condensates are given in Table 6. [16].

Table 6.

Compositions of investigated unstable condensates of different compositions, %*(mol.)*

Components	Composition		
	I	II	III
N_2	0,114	0,153	0,069
CH_4	51,326	51,634	51,088
C_2H_6	10,806	10,444	8,886

C_3H_8	9,410	7,675	1,060
iso - C_4H_{10}	2,519	2,234	2,136
n - C_4H_{10}	2,858	2,693	0,876
iso - C_5H_{12}	2,175	4,723	7,166
n - C_5H_{12}	1,869	4,257	4,934
C_6H_{14}	3,286	4,001	5,515
		Continuation of Table 5	
C_7H_{16}	2,509	3,847	5,019
C_8H_{18}	2,271	2,849	3,695
C_9H_{20}	2,218	2,155	3,852
$C_{10}H_{22}$	8,4 01	3,467	5,537
CO_2	0,239	0,226	0,667
The amount of C_5 to C_{10} in the raw material:			
% (mol.)	22,729	25,299	35,718
Kg/100 mol.	2302	2432	3548

Raw materials were treated according to three variants (Fig. 2). In the first variant stabilization was carried out in one step at 0.13 MPa and 40 °C. According to the second variant, the raw material was first degassed at 4.0 MPa and 10 °C, then the liquid phase was subjected to separation in the second stage at 0.13 MPa and 40 °C. The third variant provided stabilization of feedstock in 3 stages at the following regimes: I - 4 MPa, 10°C; II -1.6 MPa, 0°C; III - 0.13 MPa, 40°C. All variants in the closing stage of separation have the same mode in pressure and temperature.

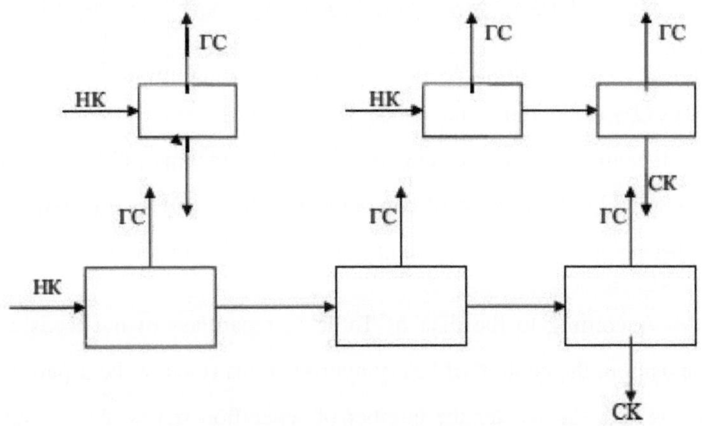

***Fig. 2.* Calculation scheme of variants of condensate stabilization by degassing.**

NK - unstable condensate; GS - separation gas; SC - stable condensate.

Table

Degree of stripping of heavy components at condensate stabilization under the variants of schemes, according to Fig. 5 and Table 6.

Components	Composition No. 1			Composition No. 2			Composition No. 3		
	I	II	III	I	II	III	I	II	III
1	2	3	4	5	6	7	8	9	10
iso - C_5	86,70	97,49	57,64	90,80	2,55	63,59	82,94	56,72	41,73
n - C_5	84,20	62,78	52,49	88,97	65,48	55,06	79,91	51,65	36,85
C(6)	66,47	38,08	28,66	74,90	43,91	34,01	59,56	28,06	17,53
C(7)	42,40	18,60	12,99	52,69	22,49	16,03	35,45	12,64	7,30
C(8)	22,30	8,11	5,49	30,08	10,04	6,84	17,46	5,27	2,94
C(9)	11,49	3,84	3,54	16,22	4,78	3,19	8,70	2,44	1,35
C(10)	5,17	1,66	1,10	7,49	2,05	1,36	3,827	1,05	0,58
1	2	3	4	5	6	7	8	9	10
L	15,417	19,49	21,04	10,95	16,81	18,86	19,98	27,55	30,83

G	1865	2164	2252	1202	1670	1837	2229	2833	30580
S	18,99	5,96	2,17	50,6	30,1	24,50	37,13	20,16	13,8

Note: *L* - content of stable condensate, % (mol.) to the amount of C_5 - C_{10} in the raw material according to Table $_5$; *G* - amount of C_5H_{12} in stable condensate, kg; *S* - amount of components $C_5H_{(12)}$- $C_{10}H_{12}$, transferred to weathering gases, % of the content in the initial raw material.

According to the data of Table 6, regardless of the feedstock composition, the content of heavy hydrocarbons (C_{5+}) in the separation gases is less, the greater the number of separation stages. At the same time, the light fraction of condensate is mainly carried away with degassing gases, which leads to a decrease in the yield of gasoline fractions when processing stable condensate.

In all variants the separation gases of different stages contain a lot of light components and do not meet the requirements for liquefied gases of GOST.

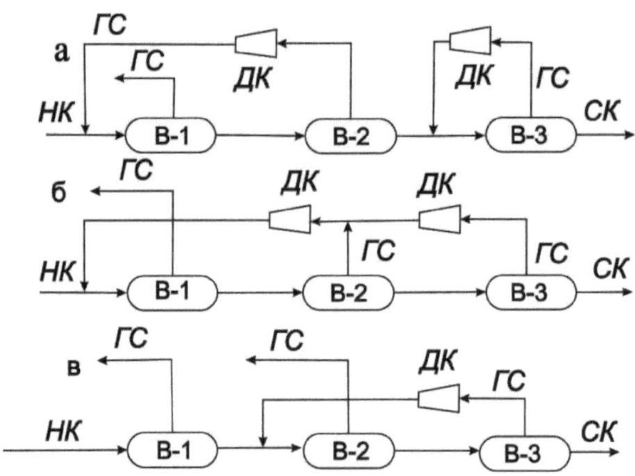

Figure 3. Principal process flow diagram of USK with recirculation of degassing gases:

B-1, B-2, B-Z degassers; DC booster compressor; NC - unstable condensate; GS - separation gas; SC - stable condensate.

Along with the above scheme, it is also possible to stabilize condensate by recirculation of separation gases into the liquid phase using a booster compressor (Fig. 3). In this case, due to the shift of equilibrium between the phases, there is an additional release of light hydrocarbons from the liquid phase. At the same time absorption of heavy components by liquid hydrocarbons also takes place. As a result the yield of stable condensate from the processed feedstock increases.

A qualitative characteristic of the operation of the above condensate stabilization schemes is given by the graphs of Fig. 4. In obtaining these dependences, the feedstock of the following composition was treated: CH_4 - *30.23;* C_2H_6 - *8.11;* C_3H_8 *-7.45; iso -* $C_{(4)}H_{10}$ *- 1.44; n -* C_4H_{10}*- 4.42; iso -* C_5H_{12} *-1.99; n -* C_5H_{12}*- 2.59;* C_6H_{14} *- 4.27;* C_7H_{16} *-39.50% (mol.).*

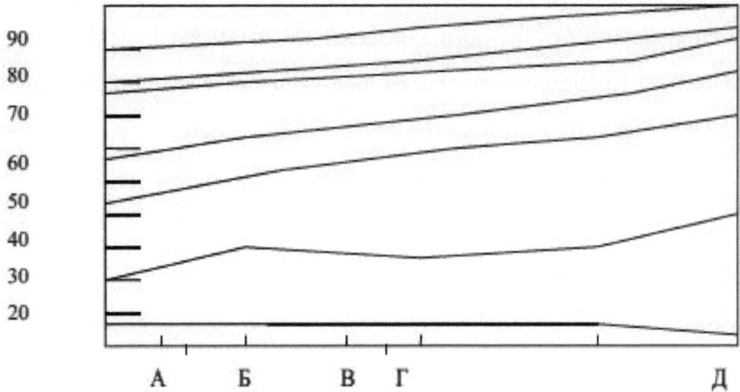

Fig. 4. Yield of components in stable condensate at different variants of USK operation [16].

Points *A, B, C, D* and *D* in Fig. 4. characterize the following variants of the separation process: *A* - 2-stage degassing of unstable condensate without recirculation; *B* - 3-stage degassing without recirculation; *C* - 3-stage degassing according to Fig. 4, *a*; *D* - 3-stage degassing according to Fig. 4, *b*; *D* - 7 - step degassing without recirculation.

The graphs of Fig. 4 show that when degassing gases are fed into the unstable condensate stream, the depth of light hydrocarbons separation from the liquid phase increases, the content of butane and heavier hydrocarbons increases.

Application of condensate stabilization in Fig. 3 requires inclusion of a booster compressor in the scheme, which increases their energy and metal consumption.

It is considered expedient in the presence of sufficiently large overpressure in the system to use in the scheme of ejectors of the "liquid - gas" type, where unstable condensate could serve as an active flow.

The advantage of the condensate stabilization scheme with multistage degassing is low metal and energy intensity and simplicity. The latter determines the normal operation of the technological scheme with the use of a minimum number of measuring devices, automatics and changes in the composition of the processed feedstock.

The process of condensate stabilization by multistage degassing has found wide application at the fields of our country having low condensate factor. Under these conditions, degassing gases of the first stage are fed into the marketable gas stream with the help of a compressor. Gases of the subsequent stages, depending on specific production conditions, are used in the fuel network or fed to the flare. The latter leads to a decrease in the efficiency of UGC operation.

Condensate stabilization by multi-stage degassing is also used as a backup option in case of shutdown of USSs operating in rectification mode.

As mentioned above, the process of condensate stabilization by multi-stage degassing has serious drawbacks, such as loss of light fractions of condensate and impossibility of production of liquefied gases meeting GOST requirements. In addition, collection and utilization of separation gases are associated with high energy costs. These factors, as well as the increase in the volume of condensate production, led to the development and introduction of new technological processes of condensate stabilization - using rectification columns. These processes have the following advantages over stabilization by multistage degassing:

pre-separation and de-ethanization of unstable condensate at high pressures facilitates utilization of gas streams;

it is possible to produce liquefied gases meeting GOST requirements without the use of artificial refrigeration;

unstable condensate energy is used rationally;

commercial condensate is characterized by low saturated vapor pressure, which reduces its losses during transportation and storage.

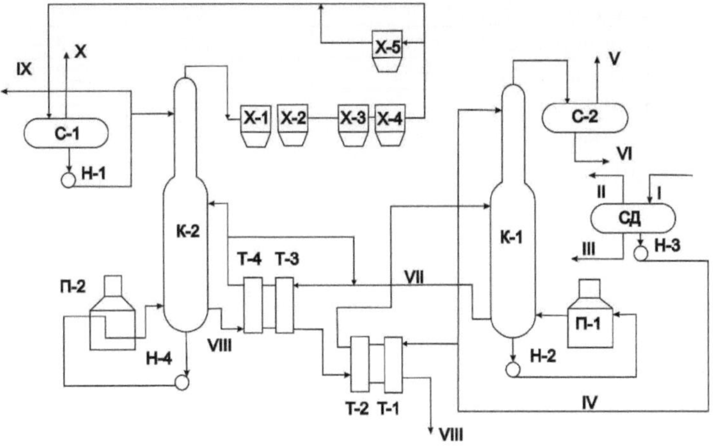

Figure 5. Technological scheme of the Shurtan gas processing plant:

C-1, C-2, SD - separators - separators; X-1, X-2, X-3, X-4, X-5 - air cooling apparatuses; T-1, T-2, T-3, T-4 - recuperative heat exchangers; P-1, P-2 - furnaces; K-1 - deethanizer; K-2 - debutanizer; H-1, H-2, H-3, H-4 - pumps; I - *unstable* condensate; *II, V, X* - degassing gas; *II, VI* - water-methanol mixture; *IV* - degassed unstable condensate; *VII* - de-ethanized condensate; *VIII* - stable condensate; *IX* - NGL.

The first USC, where the rectification process is used to obtain marketable condensate, was commissioned at the Shurtan GPP (Fig. 5.) [17].

The complex consists of 2 identical process lines including deethanization and debutanization units. The mode and main characteristics of the USK equipment are given in Table 7.
The feedstock for the unit is partially degassed unstable condensate produced at the NTS units of the Vuktylskoye gas condensate field.

Table 7.

Main indicators of USK rectification columns.

Indicators	Columns	
	K - 1	K - 2
Raw material throughput, m^3/h	300	200
Diameter of upper section, *m*	1,8	1,8
Diameter of the lower section, *m*	2,8	2,8
Column height, *m*	36	36
Number of plates in the upper section	14	14
Number of plates in the bottom section	26	26
Plate type	Two-stage valve	
Catering plates (counting from the bottom)	14	14
Pressure, *MPa*	2,1	2,1
Temperature, ^{0}C		
tops catering bottom	50 22 160	75 110 190

Unstable condensate from the field enters the inlet separator C-1 (Fig. 5.), where it is partially degassed at 1.6-1.7 MPa and 0-10°C. At the same time, the methanol-water mixture is settled, which is discharged from the system.

The feedstock is fed into the deethanizer in two streams: ~60% (wt.) is heated in the heat exchanger T-1 to 10-30 °C and introduced into the column through the 14th plate, and the second part is fed to the 22nd plate as irrigation.

The temperature at the bottom of the deethanizer is maintained by forced circulation of a part of the cube liquid through the flameless combustion furnace P-1.

The lower product of column K-1 is fed to stabilizer K-2, where it is debutanized. The vapor-gas mixture discharged from the top of the column K-2 is cooled in air condensers-coolers to 40-60 ^{0}C and enters the tank C-1. This product composition corresponds to the broad fraction of hydrocarbons (HFC) and is used to produce liquefied gases of various grades. The cube product of column K-2 corresponds to stable condensate with saturated vapor pressure not more than 66.65 kPa.

Average compositions of condensate stabilization unit flows at the Shurtan GPP are given in Table 8.

Table 8.

Average composition and number of USC flows at the Shurtan GPP

Flows	Content, % *(wt.)*						Yield from raw materials, % *(wt.)*
	N_2+CH_4	C_2H_6	C_3H_8	C_4H_{10}	C_5H_{12}	C_{6+}	
Unstable condensate	2,62	5,97	12,70	13,17	2,11	63,43	100
Separation gas	53,17	27,65	13,47	4,40	1,09	0,20	2,17
Stabilization gas	16,92	59,43	20,55	2,78	0,32	-	8,67
NGLS	-	1,42	44,97	52,91	0,70	-	18,57
Stable condensate	-	0,01	0,06	2,39	1,71	95,83	69,11

To cool the stable condensate and the upper stabilization product, air cooling units (ACU) of the AVZ-14, 6-25-B1-VZT/8-4-6 type were installed at the first and second USK in 1980 [19]. Each AHE consists of six sections, the area of external and internal heat exchange surface of each section is equal to 1250 and 65 m^2 respectively. Overall dimensions of the apparatuses: length 6650 mm, width 6230 mm, height 5864 mm. The mass of the apparatus is 3965 kg. The motor of AHE has a rotation frequency of 250 revolutions per minute. The heat transfer coefficient of air-cooled apparatuses is 110-160 W/(m^2deg). The temperature of stable condensate due to cooling in AHE is reduced to 30-40 in summer and 12-20 ^{0}C in winter

As mentioned above, separation of water-methanol mixture from the feed is performed in the inlet separator C-1. The residual content of water-methanol mixture in the unstable condensate at the inlet to the unit is up to several percent of the feedstock. Mineralization of water entering the column together with unstable condensate is 50-55 g/l. During operation of the USK there is periodic accumulation of salts in the deethanizer, especially in its lower part and in the furnace tubes. This leads to decrease of heat and mass transfer in corresponding parts of the column.

To combat salt deposits in the deethanizer, it is practiced to supply acute water steam with pressure of 3.26 MPa and temperature of 250 °C under the bottom plate of the column. Water steam is supplied if necessary during 4-5 hours [18].

Water vapor together with hydrocarbon vapor rises to the top of the column. Reaching the upper plates, water vapor condenses, water flows to the bottom of the column and is removed from the cube.

The process is monitored by the salt content in the aqueous condensate discharged from the bottom of the column. In the initial period the salt concentration in the aqueous condensate is 450 g/l, then gradually decreases and by the end of the cycle is only a few grams per liter of solution.[19].

It should be noted that in the presence of water vapor in the system, the partial pressure of light components released from the condensate decreases, which, in turn, provides a reduction in energy consumption in the deethanization unit.

To reduce energy costs, it was tried to feed hot stripping gas into the deethanizer cube [20].

Fig. 6 shows the curves of dependence of the required heat load on the AOC furnace on the amount of stripping gas. During the experiments, the stripping gas was heated to a temperature close to the column bottom temperature. The maximum reduction of heat load on the furnace of column K-1 amounted to 22%. Further increase in the amount of stripping gas leads to violation of stability of column plates operation.

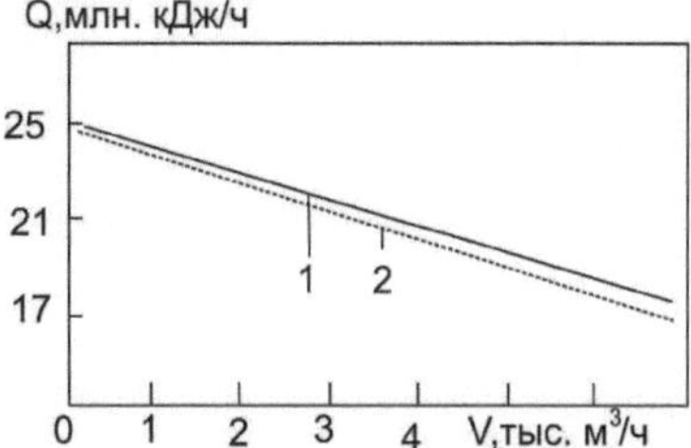

Fig. 6. Dependence of heat load on the furnace Q of column K-1 on the flow rate of stripping gas *V*. *1* - calculated; *2* - experimental

The analysis of technical and economic indicators of CCS operating under the schemes of multistage degassing and rectification was carried out for the CCS of the Shurtan gas processing plant [17]. According to the initial project it was envisaged to carry out the process of condensate stabilization by 3-stage degassing. To extract propane and higher hydrocarbons from separation gases, it was envisaged their compression with subsequent cooling and condensation. The unit produced dry gas, stable condensate and NGL as marketable products.

In 1973, the design scheme was reconstructed to a rectification scheme (Fig. 6). Condensate stabilization using the rectification process allowed to reduce condensate losses with low-pressure gases almost 3 times.

The advantage of the described scheme is its low metal and energy intensity. The unit provides, at minimal cost, the production of liquefied gas and stable condensate meeting the requirements of the relevant standards. The plant scheme is also characterized by deep heat recovery of stable condensate.

The disadvantages of this scheme are:

(a) Low degree of propane recovery from unstable condensate (~70-75%). Hydrocarbon C_{3+} entrainment with degassing and de-ethanization gases is 80-100 g/m^3;

b) strict dependence of USC operation on the performance of the CCU. This is especially noticeable when the unstable condensate flow temperature rises;

c) in case of high gas saturation of unstable condensate used as irrigation, foaming in the system is possible, which leads to carrying away of heavy hydrocarbons by deethanization gas (in the form of mist and dripping liquid);

As mentioned above, one of the main features of GPU and GPP operation is the change in the composition and quantity of feedstock over time (with a decrease in reservoir pressure of the field). For example, after seven years of operation the yield of unstable condensate at the Vuktylskoye gas condensate field decreased by 2 times. Consequently, the USC was operating at half of its design capacity. In order to maintain technical and economic performance (TEP) of the USK at a good level, feedstock from other fields can be fed to the USK. In the absence of such an opportunity, the optimal FEA can be achieved by the following methods:

(a) By changing the design scheme of the USC. According to the project, approximately two thirds of the feedstock passes the recuperative heat exchanger and enters the middle part of the column, and the rest of the feedstock as a cold spray, is fed to its upper plate.

It is obvious, however, that during the period of feedstock quantity reduction the whole unstable condensate stream can be fed to the deethanizer by a single stream (through the recuperative heat exchanger). In this case a part of stable condensate pre-cooled by raw material stream can be used as deethanizer irrigation in . When operating under this scheme, the operating costs of the USC will slightly increase, since a corresponding heat consumption will be required to cool and regenerate the stable condensate used as an absorbent. However, this will not be associated with the introduction of additional capacities for heat supply to the USS, since circulation of stable condensate (absorbent) in the system will be provided mainly by utilization of underloaded USS capacity.

By irrigating the column with stable condensate, the amount of propane carried away with the deethanization gas can be reduced several times;

b) transition to absorption technology. In cases when the amount of feedstock at the USK is reduced by 2 or more times, part of its technological strings (at the Shurtan GPP - one string) can be operated as an absorption unit for processing degassing and de-ethanization gases. In this case column K-1 (see Fig. 5) should operate in AOC mode. The operating mode of the second column will be the same as in the design scheme.

The efficiency of such USCs can be improved by supplying cold spray, under-saturated with light hydrocarbons, to the upper plate of the deethanizer. Cold spray can be obtained by cooling the deethanization gas in an air-cooled apparatus (ACO) or propane refrigeration unit. Cooling of the deethanization gas with unstable condensate from the low-temperature separation stage of the NTS unit is also used to obtain cold sprinkling.

When the USK is operating in the design mode, unstable condensate is delivered to the USK at low temperatures, so the weathering gas has a temperature 20-25 °C lower than the deethanization gas. Consequently, degassing gases can also be used to cool deethanization gases and condense a part of its heavy hydrocarbons;

c) in case of high gas saturation of irrigation of the deethanizer of the USK deethanizer operating according to the scheme of Fig. 5, stable condensate can be fed into the irrigation stream. 5 stable condensate can be fed into the irrigation stream. The amount of the latter is chosen so that the resulting mixture is undersaturated with respect to light components (C_1-C_4). It is also possible to use stable condensate as a second irrigation of the deethanizer.

3. Performance of stable gas condensates and their processing into motor fuels.

With the increase of natural gas production in Uzbekistan, the number of processing facilities such as the Mubarek GPP, the Gazli GPP, the Shurtangaz GPP, the Uchkyr JU, etc. is increasing: Mubarek GPP, Gazli GPP, Shurtangaz, Uchkyr JU, etc. The gas industry is well developed in the republic, including drilling of gas wells, extraction, processing and transportation of natural gas and gas condensate (GC.). However, petrochemistry is underdeveloped (PO "Ferganaorgsintez"), the gas chemical industry is represented only by a methane conversion shop with subsequent production of ammonia and urea at Chirchik PO "Elektrokhimprom", and this is an integral part of the integrated use of natural gas and its associated components (ethane, propane, butane, HC, helium mercaptans, etc.). Due to the lack of chemical processing, a significant share of gases, light hydrocarbons and acid gases is flared, which is unacceptable from the environmental point of view.

The development of the gas chemical industry is characterized by constantly growing consumption of hydrocarbon raw materials, so necessary for the production of, in particular, polyethylene, polypropylene and polyvinyl chloride, as proved by the world production experience.

By the level of utilization of light hydrocarbons (C_1 - C_5), the CIS countries as a whole are significantly behind the advanced foreign countries. In our country this lag significantly hinders the development of a number of sectors of the national economy. National economic prospects depend, in particular, on energy and chemicalization, directly related to the availability and efficient use of organic and mineral raw materials. To meet the needs and development of individual sectors of the national economy of the country, of great practical interest is the oxidative transformation of gas (CH_4) into formaldehyde with subsequent production of hexamethylenetetramine and methanol. The production of urea (urea) and the most valuable intermediate melamine also directly depends on deep chemical processing of gas.

The most valuable raw material for organic synthesis is HC, a companion of natural gas (T = 308-338 K), which is separated by separation at low temperatures. At the same time HC is condensed at almost all stages (collection, treatment, low-temperature separation and desulfurization) of natural gas processing (Scheme 1).

Scheme 1: Natural gas processing

At present, more than 2 million tons/year of HC is accumulated in the fields of Uzbekistan, shipped to Altyaryk NP3 for processing into motor fuel under the scheme of oil refining (direct irrigation of the atmospheric rectification unit of oil). In my master's thesis I propose to turn HC into motor fuel and paint solvent directly at its separation and stabilization from gas at the USK unit (scheme 2).

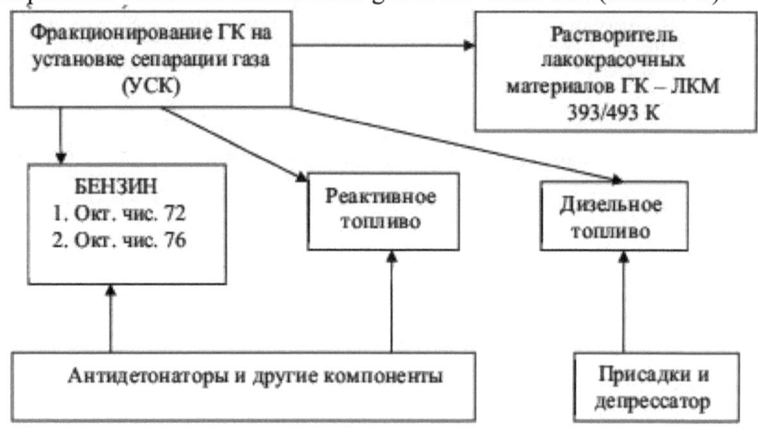

Scheme 2: Gas separation and stabilization at the USK unit

The annual processing of 450 thousand tons of HC at Shurtangaz GPU alone will yield 45 thousand tons of solvent and

400 thousand tons of motor fuels, respectively. On the basis of the existing scheme of CCS with the help of insignificant technological changes the most rational and effective method of HC extraction from gas and its directed hydrocarbon fractionation for chemical transformation into semi-products was developed. Judging by physicochemical characteristics (Fig.1), Shurtan HC contains a significant amount of aromatic (22-27%) and naphthene-paraffin (the rest) hydrocarbons.

On the basis of thermotechnical calculations and thermodynamics of separation of azeotropic mixture of hydrocarbons under pressure, the technological mode of rectification unit of the USK unit was determined. In addition, universal solvents of paintwork materials are also produced at the USK according to the proposed technology (Fig. 7). Here is a scheme of the process:

- condensate from the first and second separation stages of the 363-633 K fraction (stream 1) after regenerative heating in heat exchanger 1 and heating in heater 5 is evaporated once in degasser 7; vapor phase 308-393 K is fed into deethanizer 10, and liquid fraction 393-633 K is once again evaporated in deethanizer 8 with release of vapor phase fraction 393-493 K, which, having condensed in heat exchanger 1, forms (stream III) the target product of paintwork solvent (HC-LSH 393/493);

- condensate from low-temperature separator of 308-453 K fraction (flow II), regeneratively heated in heat exchangers 4 and 2, is supplied to deethanizer 10, where hydrocarbons up to C_3 are removed from it; further, in the presence of hydrogen sulfide, it is supplied to debuganizer 11, where hydrocarbons up to C_5 are removed from it; from debuganizer 11 condensate after heating in heater 3 goes for single evaporation to degasser 9, where it releases vapor phase 308-413 K, and liquid phase 413-453 K (flow IV) is the target product - solvent of paintwork materials. (GC-LCM 413-453).

Separate supply of condensate from separation stages to stabilization allows to obtain by single evaporation paintwork solvents of known type "White spirit" (413-453 K-aliphatic hydrocarbons) and Solvent (393-493 K - mixture of hydrocarbons with the content of 30-40% aromatics)[25]

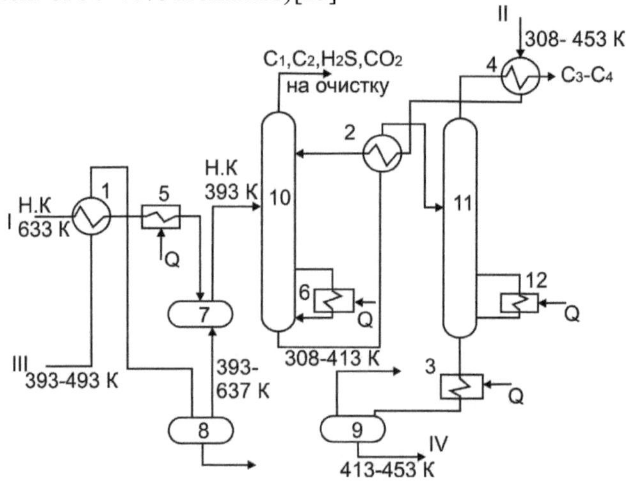

Fig. 7. Technological scheme of obtaining solvents GC - LCM 393/493 and GC - LCM 413/453.

1,2,4-heat exchangers, 5,3,6,12-heaters, 7,8,9-degassers, 10-de-ethanizer, 11-debutanizer.

By adjusting the mode of raw gas supply, pressure, temperature of vaporization and condensation of hydrocarbons, it is possible to obtain a broad fraction of light hydrocarbons (raw material for obtaining intermediates of nitriles, unsaturated compounds), a narrow fraction of hydrocarbons (product for gas-phase catalytic oxidation into oxygen-containing compounds) and heavy fractions of HC (feedstock for obtaining sulfo and oxyacids) [26]. This technology does not exclude the production of motor fuels (A-80, A-91 and summer diesel). Thus, along with the processing of HC into fuel, it is necessary to develop the field of its chemical transformation into functionally active intermediates, as well as the promising use of HC as a raw material component in the processes of reforming, cracking and pyrolysis. This makes it possible to obtain either fuel components or low-molecular olefins (ethylene, propylene and butylene) and aromatic hydrocarbons, which are raw materials for petrochemistry. Chemical processing of HCs into various intermediates and materials is also relevant due to the steady reduction of the world stock of fuel substances (oil, gas, etc.). It is irrational and mismanagement to satisfy the need for fuel at the expense of clarified natural hydrocarbon mixtures (in particular, HC).

There are recommendations for the preparation of anionic surfactants [27] and extraction of alkylbenzenes from HA [31]. Chloromegylation of alkylbenzenes in fractions for the separation of individual chloromethyl derivatives of the benzene homologous series has been studied. It is a conjugated reaction proceeding in acidic medium with electrophilic substitution of by a methylene group. After hydrochloric acid condensation, the aromatic hydrocarbons are simultaneously methylated and chlorinated. The kinetics of the reaction mechanism and the formation of isomeric benzyl chloride compounds were investigated.

On the basis of experience in chloromethylation of benzene homologues chloromethylated GC "Shurtan" containing 27.4% of aromatic, 18.8 - naphthenic, 53.8 - aliphatic hydrocarbons, the technology of obtaining "Chloromethylate" and "Concentrate" of alkylben-zyl chlorides was developed. A number of chloromethylpro-derivatives of alkylbenzenes were obtained from "Concentrate" and on their basis various surfactants were synthesized and water-soluble polymers were modified [35]. Highly effective hydrotropic substances based on HA hydrocarbons were obtained and the regularities of their influence on the colloidal-chemical properties of concentrated solutions of surfactants and synthetic detergents (CMC) were elucidated. Methods of regulating CMC properties by means of adding synthetic hydrotropic substances have also been proposed [36].

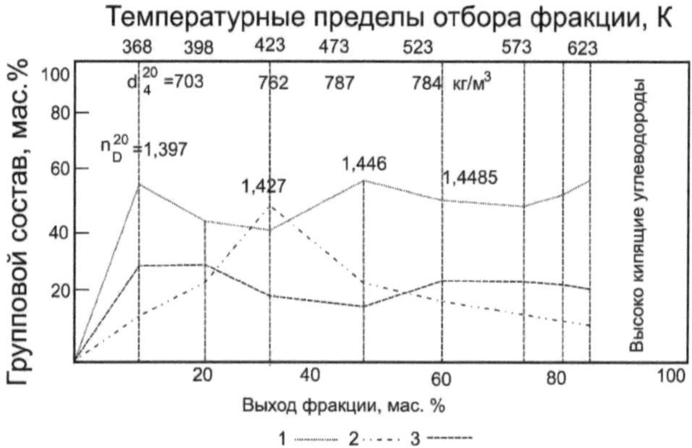

Fig. 8. Characterization of gas condensate composition of the Shurtan field.

1 - paraffinic, 2 - aromatic, 3 - naphthenic hydrocarbons.

The technology of new bisquaternary cationic surfactants from alkylbenzyl dichlorides and various tertiary amines has been synthesized and developed. Their physicochemical, surface-active properties were studied and ways of their use were proposed [37]. A surfactant with the conventional name "G-9" has been synthesized [38] from HA and studied the possibilities of its use as an inhibitor of steel corrosion. An antimicrobial additive from a mixture of aliphatic amines with amino groups at primary and secondary hydrocarbon atoms C_9 - C_{14} was synthesized. It was found that the introduction of the additive in the amount of 0.1% increases the biostability of jet fuels. It also possesses antioxidant and anticorrosive properties. Similar to hydrocarbons contained in high HC fractions, kerosene fractions of crude oil have long been used to produce sulfanol-type surfactants. In a well-known case, alkylaryl sulfonates were obtained from chlorokerosene by sulfination.

However, the above methods of chemical utilization of HA are not widely used, although some results of scientific research are of practical interest. Research in this direction is being intensively developed. They are aimed at creating a simple and rational method of processing, and the end products should be widely used in the national economy. From this point of view, the oxidizing method, in our opinion, is the most acceptable. In our experiments, the light fraction (338-448 K) was subjected to gas-phase oxidation [41], and the kerosene fraction (433-513 K) - to liquid-phase oxidation [42]. Oxidates were obtained from light hydrocarbons of HA with df=766kg/m^3, p^{20}= 1.4734 having functional compounds as well as acids (52.2 mgCON/g), alcohols (43.4) and aldehydes (21.8). The manufacturability of the process is illustrated on a model plant for the oxidation of HC hydrocarbons (Scheme 3.) [43].

Gases of deethanization, debutanization, and light fractions of HC hydrocarbons were subjected to oxidative ammonolysis to obtain intermediates for surfactants and HRP [44]. On catalysts of coupled processes containing oxides of molybdenum, nickel, titanium, cobalt and chromium on aluminum oxide carrier, under ammonia pressure of 10-20 atm. at 623-773 K, the conversion of hydrocarbons by the sum of nitrogen-containing (nitriles, amides, imides, etc.) compounds reaches 30-45 May % [45].

The given data indicate that from gases and hydrocarbons of HC at conjugate and oxidative ammonolysis, as well as chloromethylation of its alkylbenzenes are formed valuable functional-active intermediates for obtaining a number of chemical products (surfactants, surfactants, paintwork materials, solvents, plasticizers, etc.). On the basis of the given data, as well as the experience of petrochemical production is recommended as the most expedient fuel-gas-chemical scheme of complex utilization of HC (scheme 4).

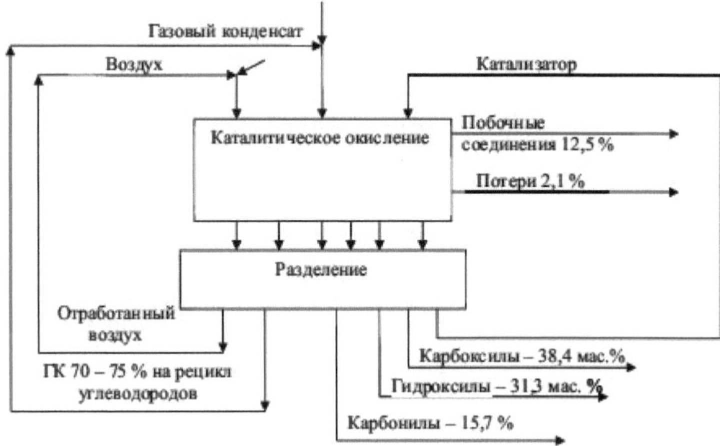

Scheme 3. Technological process is illustrated on a model plant for oxidation of hydrocarbons HC.

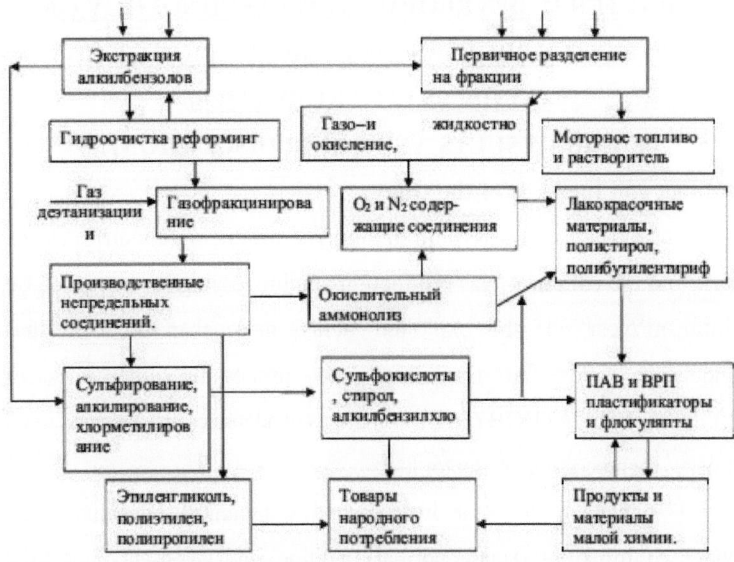

Scheme 4: Stable gas condensate.

Scheme 4 assumes the presence of separate reactions not yet developed, but the complex of issues related to creation of waste-free environmentally friendly technology allows their introduction into the process [46].

CONCLUSIONS TO CHAPTER I

Along with the fuel application of natural gas and gas condensate, it is necessary to develop the direction of their complex chemical utilization. In particular, at the Shurtan and Muborek fields, including a gas processing plant, there are all the conditions for the creation of a gas chemical complex, which will contribute not only to the significant development of the economy, but also to the improvement of the socio-ecological situation in the region.

CHAPTER II. DEVELOPMENT OF TECHNOLOGY OF EFFECTIVE PROCESS OF STABILIZATION OF GAS CONDENSATE PROPERTIES.

(MAIN RESULTS AND THEIR DISCUSSION)

1.Object and methods of the study.

Uzbekistan is the seventh among gas producing countries in the world and the volume of gas production and processing has exceeded 60 billion m^3/year. Of the exploited fields, more than 60% are gas condensate fields. At five large gas processing enterprises of UZGEONEFTEGAZDOBYCHA OJSC, gas condensates are allocated to oil refineries for processing into motor fuel.

Gas condensate is a hydrocarbon companion of natural gas, condensed from (C_1-C_4) under normal conditions with an onset boiling point of 35-45°C and an end boiling point of 360-410°C, consisting of the sum of various (C_3-C_{30}) hydrocarbons with group composition:

Paraffin	- 28-60%
Naphthenic	- 25-45%
Aromatic	- 47-37%

Physical and chemical characteristics of condensates determine their marketable properties.

To assess the possibility of obtaining separate grades of motor fuels from condensates, their unified technological classification according to the industry standard OST 51.56-79 has been established [21]. According to this classification condensates are analyzed by the following parameters: saturated vapor pressure, sulfur content, fractional composition, content of aromatic hydrocarbons and paraffins, pour point.

Condensates are subdivided into three classes according to the content of total sulfur:

I - sulfur-free and low-sulfur condensates with the mass fraction of total sulfur not exceeding 0.05 %. These condensates do not need purification from sulfur compounds;

II - sulfurous with total sulfur content from 0.05 to 0.8 %. The need for purification of condensates of this class and its distillate fractions in each specific case is decided depending on the initial requirements;

III - high-sulfur condensates with total sulfur content above 0.80 %. Inclusion of a sulfur compound treatment unit in the schemes for processing of these condensates is mandatory.

According to the mass fraction of aromatic hydrocarbons in gas condensates, they are divided into three types: A_1, A_2 and A_3. Types A_1, A_2 and A_3 include condensates containing more than 20, 15-20 and less than 15% aromatic hydrocarbons, respectively.

Gas condensates are divided into four types - H_1, H_2, H_3 and H_4 - according to the content of normal series alkane hydrocarbons in the fraction with the boiling point above 200°C and the possibility of obtaining fuel for jet engines, winter diesel fuels and liquid paraffins:

H_1 - highly paraffinic, in the fraction with boiling point 200-320 °C the content of complexing agents is not less than 25% (wt.). From these condensates it is possible to obtain liquid n-alkanes, jet and diesel fuel using the dewaxing process;

H_2 are paraffinic, the 200-320 °C fraction contains 18-25% (wt.) of complexing agents;

H_3 are low-paraffinic, the content of complexing agents in the 200-320°C fraction is 12-18% (wt.);

H_4 - paraffin-free, content of complexing agents in diesel fraction - less than 12% (wt.).

Condensates are subdivided into three groups by fractional composition - F_1, F_2 and F_3:

F_1 - condensates of light fractional composition, containing gasoline fractions not less than 80% (wt.), boiling not higher than 250 °C;

F_2 - condensates of intermediate fractional composition boiling out within the temperature range of 250-320 °C;

F_3 - condensates boiling above 320°C.

Thus, for gas condensate a technological characteristic code is established, which determines the expedient direction of its processing. For example, condensate from the Shurtanskoye field is designated by the code $IA_3N_1F_3$. The symbols included in it are decoded as follows:

I - class: content of total sulfur in condensate is not more than 0.05% (wt.); A_3 - type of condensate: content of aromatic hydrocarbons is less than 15% (wt.); H, - type: highly paraffinic condensate, in the fraction 200-320 °C the content of complexing agents is higher than 25% (wt.); F_3 - boiling end temperature is higher than 320 °C.

The object of the study is gas condensate (GC) of the Shurtan field - a clarified mixture of natural hydrocarbons accompanying natural gas (vapor-gas solution). GC differs from oil in nature, composition and appearance.

Let's consider physical and chemical properties of Shurtan gas condensate (Table 9).

Table 9.

Physical and chemical characteristics of Shurtan gas condensate

Fraction, K	Yield, %	Group composition of hydrocarbons, % wt %.			Specific gravity g/cm^3	Refractive index, η_D^{20}
		Methane	Naphtha-new	Aroma-tic		

338-363	9,3	58,4	21,8	19,8	0,637	1,3615
Continuation of Table 5						
363	26,8	41,5	32,7	25,8	0,693	1,4312
393-423	33,1	39,0	20,7	40,3	0,732	1,4327
423-448	16,6	62,0	5,5	32,3	0,753	1,4454
448-473	12,8	52,0	20,1	27,4	0,758	1,4622
473-498	8,7	49,5	21,0	28,5	0,774	1,4525

In the republic in 2022-2023, 4.5-6.0 million tons of gas condensate was produced, in subsequent years with the increase in natural gas production is expected to increase. In this regard, there is a question of a more rational and effective use of it not only as its processing into motor fuel, but also as a chemical hydrocarbon raw material.

A number of gas processing plants (Mubarek, Shurtan, Uchkir, Gazlinsk, Kokdumalak and others) are currently in operation. At these plants gas condensate is separated, stabilized and sent for processing to the refinery for motor fuel according to the oil refining scheme.

At the same time we will consider the properties of individual gas condensates of exploited gas fields.

Table 10

Characterization of gas condensates of a number of gas fields of Uzbekistan

№	Deposit	Production	Physicochemical properties	Group composition of hydrocarbons

		volume for 1998 thousand tons	η_D^{20}	g/cm3	Aromatic	Naphtha-new	Parafi-new
1	Mubarek (north)	510	1,4274	0,728	8,8	29,1	62,1
2	Kokduma-Lak	1910	1,4392	0,768	12	15	73
3		140	1,4486	0,787	24,2	36,7	39,1
4	Ghazli	60	1,4460	0,765	32	23	45
5	Mubarek (south)	210	1,4281	0,735	10	32	58
6		930	1,4417	0,762	29	22	49
Total							**3760 t.tn.**

As can be seen from Table 10, gas condensates differ in both quality and yield in terms of physical and chemical properties and group composition of hydrocarbons and fractions.

In this case, we show the true boiling point (BTP), physicochemical properties (d_4^{20} and η_D^{20}), molecular distributions (MR) from the mass selection of hydrocarbon fractions of gas condensate (Fig. 9.). As can be seen from Fig. 9, the ITC d_4^{20} η_4^{20} and MR of these fractions describe characteristic curves. It means that gas condensate of Kokdumalak field consists of (20-40% wt. %) gasoline, (25-40%) kerosene and (10-20% wt. %) diesel fuels, and also contains up to 4% wt. % of tail fractions containing - oils, asphaltenes and resinous substances.

Comparing indicators of widely exploited gas condensates of Shurtan, Mubarek and Zevardinskoye fields it is necessary to point out their fractional compatibility with Kokdumalak because different types of motor fuels are currently obtained from it.

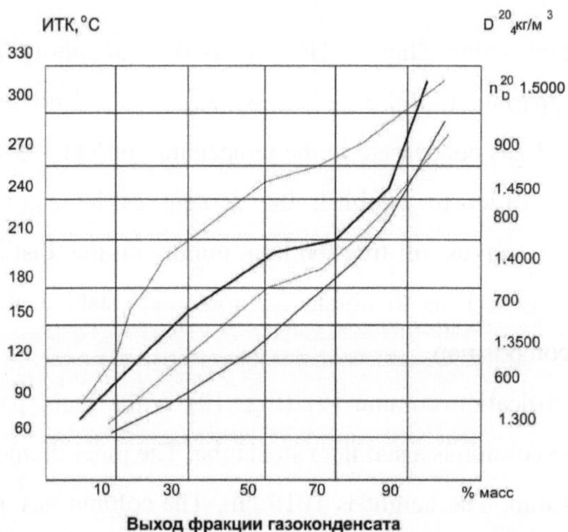

Fig. 9. *Dependence of true boiling points (TBC) and physicochemical properties on the mass of sampling fractions of Kokdumalak gas condensate.*

Properties of condensates (Table 10) were determined by standard methods, group hydrocarbon composition - by maximum aniline points, distillation was performed on an ARN-2 device. All condensates are characterized by low pour point (below-60°C) and flash point (25°C and below). The fractional composition of gas condensates is also different.

According to the group hydrocarbon composition (Table 14), the studied condensates can be divided into 2 groups: Group 1 - condensates rich in aromatic hydrocarbons (Gazli, Uchkir); Group 2 - condensates with low content of aromatic hydrocarbons and high content of paraffin hydrocarbons (South Mubarek). The content of aromatic hydrocarbons in the gasoline part of condensates of different fields varies within the following limits: Gazli - 29%; Uchkir - 34%; in condensates of other fields - 6-8%. Condensates of the second group have sharply expressed paraffin character, in their gasoline fractions 60-75% are paraffin hydrocarbons.

2. Methods for analyzing and describing the laboratory bench.

To obtain stable Shurtan HC we used a laboratory unit - oil rectification apparatus (ARN-2), which is designed for distillation of oil, oil products and gas condensate to the temperature of 743:773 K (GOST 11011-85) in order to establish the content of indicators for the construction of curves of true boiling points in the distillation of hydrocarbons, as well as to obtain fractions and establish their group hydrocarbon composition.

The rectification column (9) (Fig. 10) is the main part of the apparatus. The column is a stainless steel tube. The inner diameter of the column is 50 mm. The height is 1016 cm. The column has an electric heater on the outside. In the lower part of the column there is a grid, on which the nozzle is poured, in the beginning large (to the height of 80-110 mm), which are sections of spiral from nichrome wire with a diameter of 0.5 mm, the diameter of spiral winding is 6 mm, the height of the section is 12 mm. The rest of the column is filled with a similar nozzle with an inner diameter of spiral coiling of 3 mm and a height of 6 mm.

The temperature in the rectification column is measured using thermocouples at 3 points (top, middle, bottom). The temperature is recorded on the diagram of potentiometer KSP-2-028. To obtain solvents for mechanical engineering at the pilot plant, stable HC was weighed and poured into a cube through the throat in the amount of 3 liters and connected to the column.

Water is connected to the cooler (5) and ice is loaded into the jacket (8) of the receiver (9). Before starting rectification all taps are lubricated with vacuum grease. Manifold taps are put in positions: tap A-1,2, 4; tap B-5,7; tap B is opened, tap D is closed, tap G and clamp should be open. The tap D is closed until equilibrium is established in the column (9). The sign of equilibrium is the cessation of pressure fluctuations determined by the differential pressure gauge. After that the D tap is opened and fraction extraction starts.

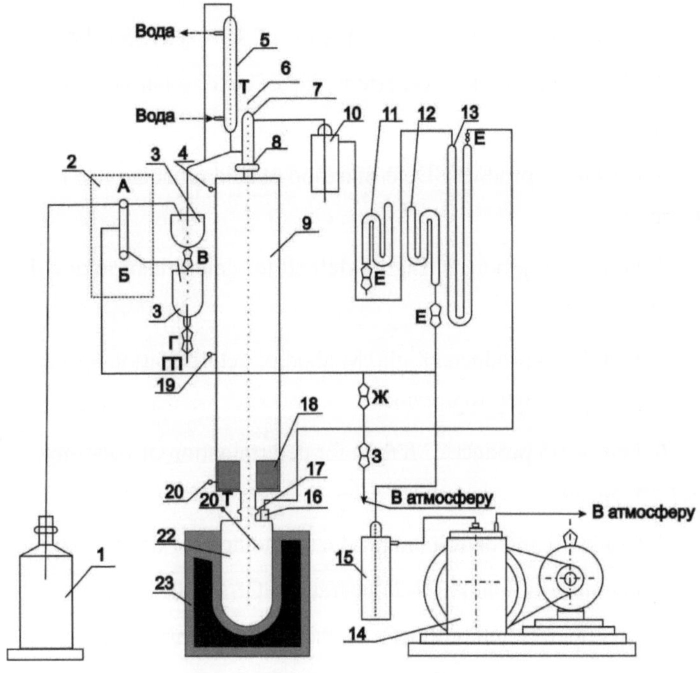

Fig. 10. Principal technological scheme of the apparatus oil rectification apparatus (ARN-2)

1-buffer tank; 2-manifold; 3-receivers; 4, 6, 19, 20, 21-thermocouples; 5-return refrigerator; 7-condenser; 8, 17-nuts; 9-rectification column; 10, 15-traps; 11, 12-mercury vacuum gauges; 13-differential manometer; 14-vacuum pump; 16-tube; 18- grate; 22- cube; 23- furnace; A- three-way tap; B- half moon tap; C, D, D, G, E- taps; 3- tap (clamp).

Rectification of stable HA was carried out at a rate of 3-4 cm^3/min, the rate was monitored with a stopwatch and measurement of distillate volume in the receivers.

Analysis and test methods for gas condensate (GC), solvents, and organodispersions were interrelated from the following standards and specifications:

1. Petroleum products. Determination of density GOST 3900-85.

2. Petroleum products. Determination of conditional viscosity GOST 6258-85.

3. petroleum products. Determination of acid number GOST 5985-79.

4. Petroleum products. Light. Method for determination of color GOST 2667-82.

5. Petroleum products. Light. Method of determination on iodometric scale GOST 19266-79.

6. Petroleum products. Method for determination of volatility GOST 12026-76.

7. Crude oil and petroleum products. Method for determination of fractional composition in ARN-2 apparatus GOST 11011-85.

8. Light petroleum products. Method for determination of aromatic hydrocarbons GOST 6994-74.

9. Petroleum products. Method for determination of flash point in closed crucible GOST 6356-75.

10. Petroleum products and hydrocarbon solvents. Method for determination of aniline point and aromatic hydrocarbons GOST 12329-77.

11. petroleum products. Method for determination of sulfur content by burning in a lamp GOST 19121-73.

12. Fuel for engines. Method of testing on copper plate. GOST 6321-69.

13. Petroleum products. Method for determining the presence of water-soluble acids and alkalis GOST 6307-75.

To determine the quality of HC, it is necessary to divide it into unstabilized (feedstock) and stabilized (chemical or fuel feedstock) [28]. When unstabilized condensates are used as feedstock for fuel and chemical production, stabilization of properties and composition - deethanization, debutanization [29] and low-temperature separation (LTS) [30], as well as absorption-desorption separation of light hydrocarbons [32] - is necessary. The scheme of separation of HC and natural gas of the field using the processes of vaporization and condensation or rectification of hydrocarbons with separation of stable HC has been mastered and operated at the facilities [33]

Table 11.

Physicochemical characteristics of local gas condensate fractions presented below

Fraction, GC, °K	Fraction yield at gas end, %	Group composition of hydrocarbons of Kokdumalak GC, % wt %.			Specific d_4^{20}	Refractive index n^{20}_D
		meta-new	naphthenic	aromatic		
338-363	9,3	58,4	21,8	19,8	714,0	1,4215
363-393	26,8	41,5	32,7	25,8	723,2	1,4312
393-423	33,1	39,0	20,7	40,3	752,5	1,4327
423-448	16,6	62,0	5,5	32,3	753,8	1,4454
448-473	12,8	52,0	20,1	27,9	758,4	1,4622
473-498	8,7	49,5	21,0	28,5	774,1	1,4525
Indicators of fraction of Gazlinsky GC						
338-448	74,8	50,4	21,8	27,8	739,0	1,4385
433-513	20,6	59,1	19,5	21,8	768,7	1,4694

Indicators of Shurtan GC						
363-448	41,2	51,2	21,1	27,7	741,1	1,4394
448-513	17,8	52,2	19,1	28,7	773,5	1,4715

Prem..: *-The residue in the flask and distillation losses were not considered here.

It follows from the above that HC with insignificant purification and fractionation meets the requirements for aliphatic hydrocarbons used in the production of fuels as their feedstock.

Table 12 outlines the parameters of the stable gas condensate after stabilization.

Table 12

Fraction, ^0C	Fraction yield at gas end, %	Group composition of hydrocarbons, % wt %.			Specific gravity kg/m(3 D_4^{20}	Refractive index n^{20}_D
		meta-new	naphthenic	aromatic		
65-90	9,3	58,4	21,8	19,8	637,1	1,3615
90-120	26,8	41,5	32,7	28,8	639,1	1,4314
120-150	33,1	39,0	20,7	40,3	732,5	1,4327
150-175	16,6	62,0	5,5	32,3	733,8	1,4454
175-200	12,8	52,0	20,0	27,0	758,4	1,4622
200-225	8,7	49,5	21,0	28,5	774,0	1,4525

Group composition of hydrocarbons in the Shurtan gas condensate field (Table 13).

Table 13

№	Indicator	NK-95 °S	95-125 °C	125-150 °C	150-200 °C	NK-200 °C
1	Gas condensate outlet	9,2	11,4	10,1	16,1	46,7

2	Refractive index, n^{20}	1,3970	1,4270	1,4460	1,4412	1,4315
3	Density d_{20}	0,7031	0,7625	0,7876	0,7836	0,7647
4	-aromatic	12,6	27,0	43,7	23,8	26,6
5	-naphthenic	29,0	31,5	17,3	14,0	20,6
6	-paraffinic	58,4	41,5	39,0	62,2	52,0
Of which:						
7	Normal structure	22,3	17,1	13,4	25,1	20,1
8	Isomeric structure	36,1	24,4	25,6	37,1	31,8

The most convenient semi-product for straight-run gasoline production is gas condensate from the Shurtan and Mubarak natural gas fields.[34] The most convenient semi-product for straight-run gasoline production is gas condensate from the Shurtan and Mubarak natural gas fields.

3. technology of stabilization of gas condensate raw materials

Sulfur condensates are stabilized according to the schemes similar to the schemes of sulfur-free condensate stabilization units. The difference between the schemes of sulfur-free and sulfur-containing condensates stabilization units is in their hardware design and regime parameters. In addition, when stabilizing sulfur condensates, separate units of the unit should be inhibited for corrosion control.

Currently, the largest sulfur condensate stabilization units are operated at the Mubarek and Shurtan UDPPs.[36-37]

The stabilization units of the USK include pre-degassing of feedstock followed by its stabilization in a rectification column. . The main differences between stabilization units at different stages of the GPP relate to processing of gas streams separated from unstable condensate.

Analysis of operation of USK-1. Preliminary demethanization of condensate is performed in apparatus B01 (Fig. 10 a). Separation gases are combined with degassing gases of amine solutions from desulfurization units and flow in one stream to absorber C02 for purification from acidic components. The purified gas is used in the fuel network.

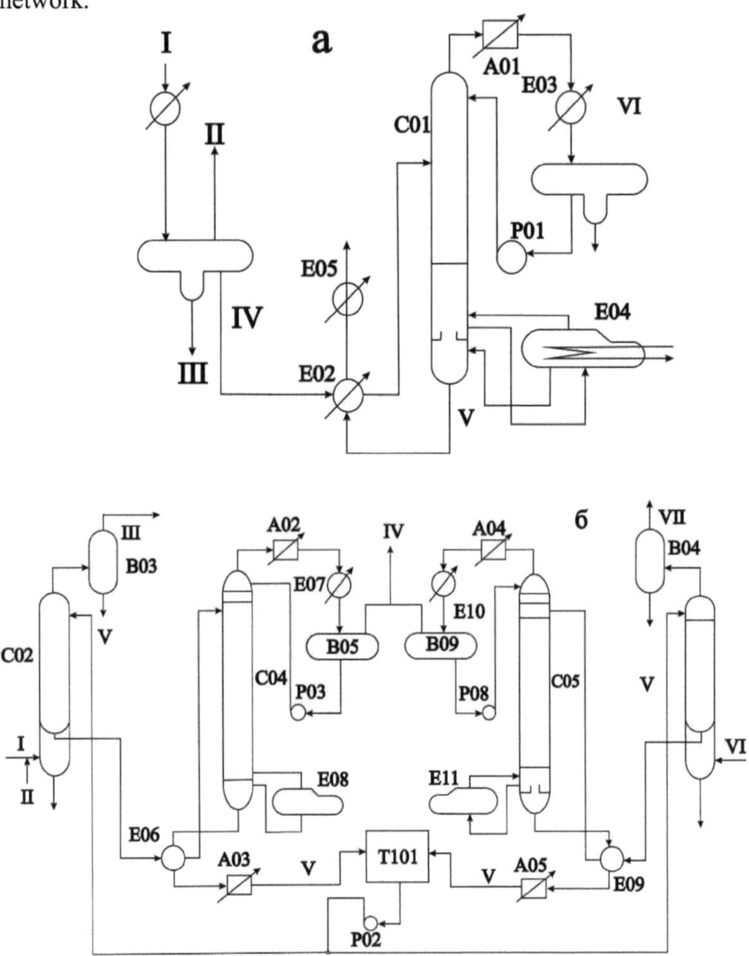

Figure 10. Principal circuit diagram of the USK-1:

(a) Stabilization unit:

C01 - debutanizer, B01 - three-phase splitter; B02 - irrigation tank; E01 - refrigerator, E02 - regenerative heat exchanger; EOZ, E05 - water coolers; A01 - air cooler, E04 - evaporator, P01 - pump, I - unstable condensate; II - degassing gas; III - acid water, IV - condensate, V - stable condensate; VI - stabilization gas.

b) stabilization gas purification unit:

C02, POP - absorbers, C04, C05 - desorbers, WHO, B04 - separators; B05, B09 - irrigation tanks; A02, AOZ, A04, A05 - air cooling apparatuses, E07, EY - water coolers, EOS, E09 - regenerative heat exchanger; E08, EP-evaporators, T101 - collecting tank, P02, ROZ, P08 - pumps, I - degassing gas to B01 (Fig. 10, a), II - expansion gas of saturated amine solution; III - fuel gas; IV - acid gases to gas sulfur production unit, V - regenerated amine solution, VI - stabilization gas to purification from B02 (see Fig. 10, a); VII - purified stabilization gas for processing.

Table 14

Characteristics of separators, degassers and irrigation tanks USK-1

Position in Fig. 10	Estimated parameters.		Diameter, m	Height (length), m	Volume, m^3	Position in Figure 10	Estimated parameters		Diameter, m	Height (length), m	Volume, m^3
	P, MPa	*T, ^0C*					*P, MPa*	*T, ^0C*			
B01	4,59	20	3,0	11,0	69,0	B04	6,63	7	1,6	5,0	6,0
B02	6,12	12	2,4	4,6	27,9	B05	5,61	4	1,4	4,0	5,3
OZ	6,63	7	1,4	4,6	45,0	B09	5,61	4	1,4	4,0	5,3

Table 15.

Characteristics of columns USK-1

Positions according to Fig. 2.2.1	Diameter, m	Height, m	Number of plates	Plate type	Project mode			
					P, MPa	Temperature, "C		
						catering	tops	Niza
C01	3,2	24,5	19	Valve	0,76	- 10	67	167

C02	2,2	23	20	2-flow	45	50	50	59
POP	2,0	23	20		0,62	45	50	59
C04	2,2	25	21	Caps	0,12	105	10	130
C05	2,2	25	21		0,12	105	110	130

Debutanization of condensate is carried out in column C01, which has 19 double-flow valve plates. Condensate stabilization gas from the top of irrigation tank B02 is discharged for purification from acidic components.

Characteristics of the main equipment of USK-1 are given in Tables 14, 15, 16. and indicators of the stabilization unit are given in Table 17.

The units show significant fluctuations in the amount of unstable condensate processed and degassing gas yield, which is also evidenced by changes in the specific yield of stabilization gases per 1 m$^{(3)}$ of stable condensate.

Table 16.

Indicators of heat exchangers USK-1

Positions on fig. 10	Quantity	*F*, m^2	Pipe space		Intertube space		Thermal load, *million kJ/h*
			P, MPa	*T, °C*	*P, MPa*	*T, °C*	
E01	1	55,8	4,08	35	0,61	206	15,38
E02	2	372	2,04	105	1,33	205	44,83
EOZ	1	450	0,51	55	1,22	65	5,14
E04	2	321	4,48	240	1,33	205	55,56
EO5	1	212	0,51	55	1,33	70	1,85
EO6	2	292	0,71	120	0,49	135	31,67
EO7	1	67	0,51	55	0,41	75	0,63
EO8	1	515	0,61	200	0,51	140	44,92

EO9	2	292	0,51	120	0,49	145	31,67
E10	1	67	0,61	55	0,41	75	0,63
E11	1	515	0,61	200	0,51	143	44,92

To ensure complete stripping of hydrogen sulfide from condensate, the project provided for maintenance of stabilizer bottom temperature 165-170 °C. At such regime pentane content in stabilization gases was allowed about 9%. During the survey period the temperature at the bottom of the C01 column was maintained at about 140°C. This mode provides almost complete purification of condensate from hydrogen sulfide. However, butanes content in the condensate was slightly higher than the design one, besides, the commercial condensate contained up to 0.2 % of propane. Despite this, the saturated vapor pressure of stable condensate does not exceed the design level of -66.7 kPa.

Increasing the bottom temperature of the CO1 column by 10-15 degrees would ensure complete propane separation and deeper extraction of butanes from the condensate.

The experience of USK operation has shown that under poor phase separation at field gas condensate processing units with unstable condensate the unit receives mineralized water. Mineral salts are partially deposited on the surface of the apparatuses, including heat exchanger E01. This reduces the heat transfer coefficient and thus does not ensure heating of the mixture before degasser B01 up to the design temperature - 20 ^{0}C.

The second main unit of USK-1 is the degassing gas purification and condensate stabilization unit (Fig. 10,b*)*.

Pressure in absorbers and desorbers of purification units is maintained at 0.55 and 0.17 *MPa*, respectively. As an absorber of acidic components 12-18 %(wt.) (according to the project 25%) aqueous solution of diethanolamine (DEA) is used. At operation of the unit in such mode the content of hydrogen sulfide in the purified gas does not exceed 5.7 *mg/m*^{3}. Concentration of H_2S and CO_2 in degassing gases makes 3,5-4,7 and 0,5-0,6 % accordingly. Degassing gases are purified with DEA solution of 12-14% (wt.) concentration at the solution to gas ratio of 2.9 -3.5 l/m(3).

Concentration of H_2S and CO_2 in stabilization gases was approximately 2 times higher than in degassing gases and was 8.9 -11.2 and 0.6 - 1.5 % (vol.), respectively. Stabilization gases in the amount of 13-15 thousand *m*^{3}*/h* are purified with DEA solution of 18 % (wt.) concentration at the solution-to-gas ratio of 5.3-6.1 l/m^3.

Regeneration of saturated DEA solutions was carried out in desorbers at a pressure of 0.18 *MPa.* Steam consumption (0.51 *MPa*) for regeneration was 120 -130 *kg/m*^{3} of solution. Under these conditions the content of H_2S in the regenerated DEA solution did not exceed 0.01 *mol/mol,* which provided a fine purification of gas from hydrogen sulfide.

At present DEA solution is supplied to the C02 and CO3 column from desulfurization units of the plant's first stage. Production of thiols mixture was organized on the basis of the equipment of amine solution regeneration units.

Sour gases after desorbers in the amount of about 4 thousand m^3/h with H_2S and CO_2 content up to 80 and 5-11% (vol.) respectively are directed to Claus plants for elemental sulfur production.

In the majority of measurements, the degree of saturation of DEA solution was 0.5-0.6 mol/mol, which is slightly higher than the permissible level. The degree of saturation of DEA solution can be reduced by increasing the DEA concentration or increasing the amount of circulating solution. Calculations carried out by us have shown that for the absorber of degassing gas purification the optimum concentration of the solution should be about 20%. At that, the solution : gas ratio should be about 3 l/m^3 [22]. For the stabilization gas purification process it is reasonable to increase the amount of circulating absorbent up to 100-110 m^3/h and to carry out absorption at a specific absorber flow rate of 6 l/m^3.

Implementation of the above recommendations will reduce the corrosion hazard at the plant while maintaining a high degree of solution saturation with acid gases (0.4 mol/mol).

Analysis of USK-2 operation. The principal difference between the condensate stabilization unit of Stage II and the condensate stabilization unit of Stage I is the use of low-temperature condensation process for separation of NGL from stabilization gases. The use of the STC process simultaneously provides NGL drying, which increases the reliability of the transportation and storage systems.

The flow of unstable condensate from the GTU at temperatures from -10 to +40 ° C enters the inlet separator D101, where partial demethanization of raw materials takes place at 3.7-3.9 MPa.(Fig. 11.).

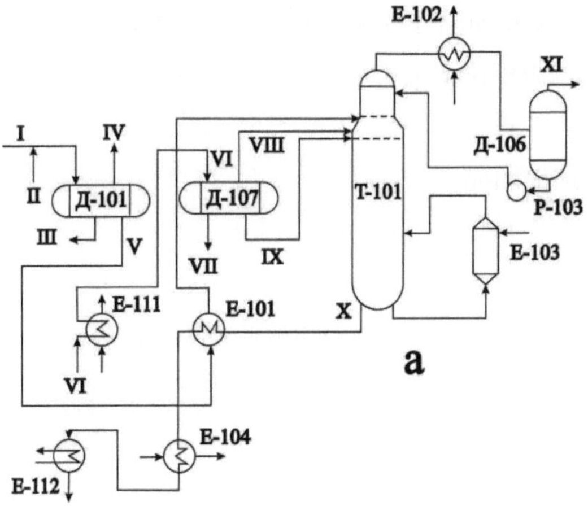

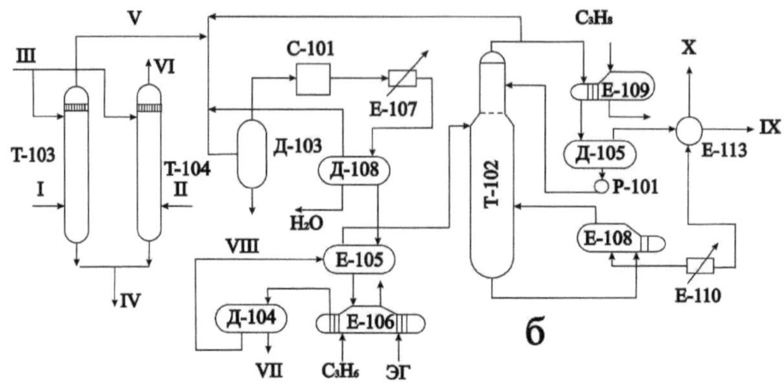

Figure 11. Principle scheme of USK-2:

(a) Stabilization unit:

D-101, D-107 - three-phase separators; D-106 - irrigation tank; E-102 - air cooler; E-101 - recuperative heat exchanger; E-103 - evaporator - T-101_stabilizer; E-111 - heater; E-104, E-112 - coolers; P-103 - pump; T-101 - stabilizer; I - unstable condensate from the field; II - unstable condensate from drying units; III - acid water; IV - separation gas for purification; V - partially degassed condensate; VI - mixture of saturated EG and unstable condensate; VII - saturated solution of EG for regeneration; VIII - gas to stabilizer; IX - liquid hydrocarbons to stabilizer; X - stable condensate; XI - stabilization gas for purification.

b) stabilization gas purification and NGL separation unit:

T-102 - rectification column; T-103, T-104 - absorbers; D-103, D-104, D-108 - separators; D-105 - irrigation tank; E-105, E-113 - regenerative heat exchangers; E-106, E-109 - propane evaporators; E-107, E-110 - air cooling apparatuses; E-108 - evaporator; C-101 - compressor; P-101 - pump; I - degassing gas from D-101 (fig. 11, a); II - stabilization gas from D-106 (Fig. 11, a); III - regenerated amine solution; IV - saturated amine solution; V - purified stabilization gas -

VI, IX - fuel gas; VII - saturated MEG solution; VIII - liquid hydrocarbon mixture; X - SFLC.

Separation gas from the top of D-101 is fed to absorber T-104, where it is purified from hydrogen sulfide and carbon dioxide with the help of diethanolamine solution. The purified gas is supplied to the fuel network.

Partially evaporated condensate from the bottom of the three-phase splitter D-101 passes the recuperative heat exchanger E-101, where it is heated up to 25 °C, and enters the 6th (counting from the top) plate of the stabilizer T-101. Two streams from the three-phase splitter D-107 are also fed to the 6th stabilizer plate as feedstock.

The gas mixture obtained during condensate stabilization from the top of separator D-106 goes to absorber T-103 for purification from acidic components. Stable condensate is discharged from the bottom of column 1-101 to the product park.

Condensate at the outlet from the cooler E-112 had a high temperature, which in summer reached 50-70°C. This resulted in losses of its light fractions during storage. A water cooler was additionally included in the scheme to cool down stable condensate.

Absorbers T-103 and T-104 in 2 tiers are filled with nozzle of Pall rings type made of plastic. In the upper part of the column T-103 and T-104 two and three baffle plates and louver package are installed respectively. Wash water is fed to the first plate to capture the amine carried away with the gas. At the amine inlet to absorbers T - 103 and T - 104 the antifoaming agent solution is fed into the flow. Stabilization gas purified from acidic components passes through the separator, is separated from moisture droplets, then it is pressed to 3.65 *MPa* and for separation from moisture enters the separator D-108.

A certain volume of gas must be supplied to the compressor to ensure normal operation. It is possible to recirculate part of the gas flows from the top of column T-102 and separator D-108 when the amount of gas from the top of separator D-103 is insufficient for normal operation of compressor C-101.

Liquid phase from the bottom of the separator D-106 through the recuperative heat exchanger E-105 enters the propane vaporizer E-106, where it is cooled to minus 30-33 °C, and enters the three-phase splitter D-104. Liquid hydrocarbon phase from the bottom of D-104 passes through the recuperative heat exchanger E-105, is heated to minus 15-10°C and, having united with the gas flow *from the* top of the separator D-104, enters the stabilizer T-102.

It should be noted that the feedstock, cooled in the evaporator E-106 to minus 33 °C, enters the column at a temperature of minus 10-15 °C. Thus, the condensation of the raw material occurs first, then its re-evaporation. The purpose of deep cooling of the raw material is to dehydrate it. Due to the cooling of the feedstock, there is no need to dehydrate NGL at a separate plant.

To absorb moisture and prevent hydrate formation, 80% monoethylene glycol (MEG) solution is injected into the tube grid of heat exchangers E-105 and evaporator E-106. The saturated MEG solution mixed with the precipitated condensate is discharged from separator D-104 and through heaters E-111 enters separator D-107 (Fig. 11. b).

The mass fraction of amine in saturated MEG solution reaches 5-10%, which indicates entrainment of DEA solution from desulfurization absorbers T-103 and T-104.

Table 17. shows the characteristics of separators and degassers at the Gas Condensate Stabilization Unit (GCS-2)

Table 17.

Characteristics of separators and degassers USK-2

Position according to Fig. 2.2.2	Pressure, *MPa*	Temperature, ^{0}C	Diameter, *m*	Height (length), *m*	Volume, m^3
Д-101	3,6-3,9	29	2,2	8,16	28
Д-103			2,2	5,7	12,1
Д-104	3,1-3,3	-33	2,8	13,4	84
Д-105	2,8-3,0	-32	1,6	5,1	9,24
Д-106	8,6-10,6	50	1,7	7,4	14,8
Д-107	1,0-1,3	40	3,1	13,8	97,7
Д-108	5,0		2,2	8,4	29,1

Proceeding from the above stated We offer the modernized technology of USK, which shows the possibilities of intensification of the process of stabilization of HA and improvement of the quality of properties of supplied HA for processing at this unit (Fig.12.).

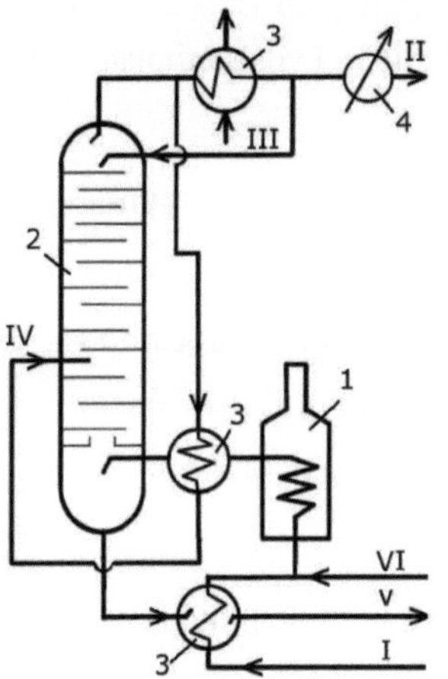

Material Flows:
I- unstable HC from the plant pre-treatment and low-temperature separation of natural gas;
II- propane-butane to propane-butane separation unit for their liquefaction;
III- light hydrocarbons for irrigation to intensify deflegmation of dissolved gases from HC;
IV- light hydrocarbon vapors for heating and use as an additional heat carrier;
V- stable HC in storage for shipment to refineries.

Fig. 12. Principal technological scheme of HC stabilization with twofold use of light hydrocarbon fractions for intensification of deflagging and as a coolant.

1- GC heating furnace, 2- rectification column, 3- heat exchangers, 4- Refrigerator.

The proposed technological scheme shows the newly introduced mode of in-line heating of incoming HC volumes and double use of light hydrocarbon phlegm (III) so that the processes of more efficient separation in the rectification column (2) of associated gases (II) and its qualitative stabilization (V). At the same time in the furnace (1) heating of HC to 240-260°C are fed NTS gases after coolers (3,4) extraction of propane-butane (P/B), which are also additional volumes for their liquefaction as fuels.

According to the proposed technology for stabilization of HC properties, a stable condensate with the following properties was obtained:

Specific gravity, kg/m^3	0,768	Methane hydrocarbons, % wt %	42,5
Refractive index, n^{20}_D	1,4575	Aromatic hydrocarbons, % wt%	24,1
Boiling point Beginning, ^{0}C,	308	Naphthenic hydrocarbons, % wt%	28,5
End, ^{0}S,	535	Mechanical impurities, % mass	absent.
		Sulfur compounds, % mass	0,25

Thus, at the above mentioned modernization of the USK and improvement of its technological mode more complete separation and increased volumes of P/B, improvement of indicators quality of stable HA, as well as significant savings in energy costs for the operation of the plant are achieved.[35].

4. studies of effective extraction of light hydrocarbons from gas condensate feedstock.

Consider a column (Fig. 13) that continuously receives F mol/h of feed mixture, which is separated into D mol/h of distillate (upper product) W mol/h of cube residue (lower product).

At steady-state, the outgoing flow is equal to the in-flow:

$$\mathbf{F\,D = + \quad W\ (1)}$$

If the concentration of the more volatile component in these three streams is equal to z_F, and, respectively, then the balance for this component is determined by Eq:

$$\mathbf{F z_F = D x_{(D)} + W x_{(W)} \qquad (2)}$$

Analysis of equations (1) and (2) shows that when F and z_F are constant and X_D and X_W meet the desired product purity, the flow rates D and W are also constant values.

Consider now the section of the apparatus bounded by the dashed line II by the n-th plate. If there is Vn mol/h of vapor rising from the n-th plate, L_{n+1} mol/h of liquid flowing from the above (n+1)-th plate, then the material balance for this section can be represented as follows:

$$\mathbf{V_n = L_{(n)+1} + \ D\ (3)}$$

If the compositions of vapor and liquid flows between the n-th and (n+1)-th plates are y_n and x_n+1, respectively, then the balance for the volatile component is as follows:

$$\mathbf{V_n y_n = L_{(n+1)} X_{n+1} + D X_{(D)} (4)}$$

Hence:

$$y = \frac{L_{n+1}}{V_n} X_{n+1} + \frac{D}{V_n} X_D \quad \mathbf{(5)}$$

Similarly, for the section bounded by the dashed line III in Fig. 2.3.1, we obtain:

$$L_{m+1} = V_{(m)} + W \quad (6)$$

и

$$y_m = \frac{L_{m+1}}{V_m} X_{m+1} - \frac{W}{V_m} X_w \quad (7)$$

where, L_{m+1}, $V_{(m)}$ - flows (in mol/h) of liquid and vapor between plates m and (m+1); $y_{(m)}$, $X_{(m)(+1)}$ - compositions of vapor and liquid flows between the same plates.

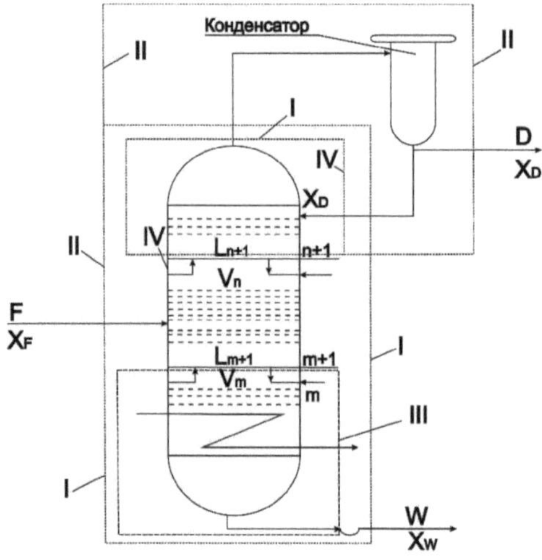

Fig. ***13. Schematic diagram of material flows in a continuously operating***

distillation column

If D and X_D are fixed based on the consideration just given, then equations (3) and (5) are insufficient to calculate the values of V_n, $L_{(n)(+1)}$, and y_n in the section of the rectification section where the liquid composition is $X_{(n)(+1)}$. As will be shown later, such a calculation additionally requires an enthalpy balance for section II in Figure 13. Similar reasoning shows that section III also requires an enthalpy balance equation to calculate the values of V_m, $L_{(m)(+1)}$, and y_n for that section in the exhaustive section where the liquid composition is $X_{(m)(+1)}$.

However, in many cases, the values of V_n, $L_{n(+1)}$, V_m, and L_{m+1} remain nearly constant for the plates, so that enthalpy balances need not be considered. Constancy of flow rates or equality of mole flows over the height of the column will be achieved under the following conditions:

1. The molar heat of vaporization values of the two components are equal;
2. The enthalpy changes with temperature are very small compared to the heat of vaporization;
3. The heat of mixing of components in both phases is zero;
4. There are no heat losses to the environment.

When the mole flows do not change, the ratio between the velocities of the flows above and below the feed plate depends on the thermal characteristic of the feed mixture. If the feed liquid mixture is fed to the column at boiling point, then

$$\mathbf{L_{m+1} = L_{(n)+1} \text{ and } V_m = N_{(n)} \; (8)}$$

If saturated steam is supplied as a power supply, then

$$\mathbf{L_{m+1} = L_{n+1} \text{ and } V_{(m)} = V_{(n)} - F \; (9)}$$

The graphical method of McCabe-Tiele [38] can be applied to determine the number of theoretical plates or contact stages required for a given distillation process of a binary mixture. Taking the molar fluxes equal, the material balance equations (5) and (7) can be easily represented graphically in the form of straight lines: y values are plotted along the ordinate and x values along the abscissa (Fig. 14.). Such lines are called working lines. Their slope is equal to the ratio of molar velocities of liquid and vapor flows. The same graph is used to plot the equilibrium relationship between the composition of vapor and liquid for the mixture of interest at the selected pressure. We need to separate 2.5 mol/h of a hydrocarbon mixture containing 1 mol/h gasoline and 0.8 mol/h light gasoline. The separation is carried out in a continuous plate column at a total pressure of 1atm. The desired final compositions, expressed in terms of mole fractions, are: X_D= 0.96 and X_W = 0.04; saturated steam is used as feed, with complete condensation of vapors in the condenser. Assuming that the phlegm flux $\mathbf{L_{n+1} = 4\,D}$, we find the number of plates required.

First, let us determine W and D. By comparing equations (1) and (2) and substituting the known values, we obtain:

$$\mathbf{Fz_F = Dx_{(D)} + (F-D)x_W}$$

$$\mathbf{4.3 \times 2.5 = D\,0.96 + (4.3 - D)\,0.04}$$

Hence D = 11.1 mol/h. According to equation (1), W = F - D = 4.3 - 11.1 = 6.8 mol/h. Next, let's calculate the internal fluxes. By equation (3), $V_n = L_{(n)(+1)}\,D +$ = 4D + D = 44.4 + 11.1 = 55.5 mol/h. By equation (9) $L_{m+1} = L_{(n)+1}$ = 55.5 mol/h; $V_{(m)} = V_{(n)}$ - F = 55.5 - 4.3 = 51.2 mol/h.

Substituting the known values into equations (5) and (7), we obtain the equations of the working line. For the rectification section (upper reinforcing part of the column):

$$y_n = \frac{44,4}{55,5} X_{n+1} + \frac{11,1}{55,5} 0,96$$

$$y_n = 0.800\, X_{n+1} + 0.192$$

For the stripping section (bottom of the column):

$$y_m = \frac{44,4}{51,2} X_{m+1} + \frac{6,8}{51,2} 0,04$$

$$y_m = 0.86\, X_{m+1} - 0.0053$$

Both equations of working lines are plotted on the y - x diagram as shown in Fig. 14. This diagram is characterized by the fact that the working line of the rectification section crosses the diagonal at the point $x = X_D$, and the working line of the exhaustive section at the point $X = X_W$.

The calculated deviation in the number of theoretical plates is due to the fact that the adopted hydrocarbon mixture for rectification boiling is very different from binary and ternary mixtures. The figure shows that the number of working column plates is not more than 30 with 80 theoretical and 40 practical column plates, and the phlegm number is not more than 3.

Hydrocarbon content in liquid X, mol. fraction

Fig. 14: Solving the example by the graphical method of McCabe-Tiele

CONCLUSIONS TO CHAPTER II

Our proposed technology of modernization of gas condensate stabilization unit and improvement of its technological mode achieves a more complete separation of fractions and increase the volume of light hydrocarbons such as propane and butane fraction, a significant improvement in the quality of stable GC, the use of propane-butane fraction as a fuel for internal combustion engines, and significant savings in energy costs for the operation of the unit.

CHAPTER III. QUALITATIVE INDICATORS OF GAS CONDENSATE

1. Qualitative-quantitative indicators characterizing stable gas condensate.

Hydrocarbon liquefied fuel gases for municipal and household consumption according to GOST 20448-80 have the following grades: SZTPB - winter technical propane and butane mixture; SLTPB - summer technical propane and butane mixture; BT - technical butane. Basic quality requirements for liquefied gases.

Previously, liquefied gases were produced according to GOST 10196-62, according to which the content of propane and propylene in liquefied gases should be not less than 93% (wt.). Such fuel has much better performance properties than liquefied gases produced in accordance with GOST 20448-75.

However, GOST 10196-62 did not stimulate the use of butanes and butylenes in the composition of liquefied gases. As a result, many gas and oil refineries used the available resources of butanes and butylenes irrationally. Introduction of the new GOST for liquefied gas allowed to increase the efficiency of C_4 hydrocarbon resources utilization and to increase the output of marketable products.

The PBP fraction is used as pyrolysis feedstock at the Uchkyr Organic Products Plant. The concentration of the components in PBP is set solely based on the requirements of pyrolysis production. In accordance with the specifications, the total content of propane and butanes in PBP should be not less than 90% (wt.), including isobutane not less than 17%, and the content of ethane and C_5H_{12+} is not more than 3 and 7% (wt.), respectively.

The ethane concentration in NGLs and LPGs is set so as to ensure their marketability and minimize losses during storage and transportation. The latter is directly related to their ethane content. Consequently, a product that does not contain ethane would have the best marketability. However, the production of NGLs and ethane-free liquefied gases is energy intensive. In view of this fact, optimal norms for ethane content in these products have been established.

The content of pentane and higher hydrocarbons in the liquefied gases is set so that they can be vaporized when the liquefied gases are used as fuel.

Table 18

Basic quality requirements for stable condensate of groups I and II.

Indicators	I	II	Test methods
Boiling point, °C, not lower	30	30	GOST 2177-99
Saturated vapor pressure, Pa (mm Hg): summer period, no more Winter period, not more	66661 (500) 93325 (700)	66661 (500) 933325 (700)	GOST
Mass fraction of water, %, max.	0,03	0,5	GOST 2477-65
Mass fraction of mechanical impurities, %, max.	0,005	0,1	GOST 6370-83
Mass of chlorides, mg/l, not more than	15	-	GOST 21534-76
Density at 20 ° C, g/cm^3	Not rationed, definition is mandatory		GOST 3900-85
Mass fraction of total sulfur, %	Same		GOST 19121 -73

According to OST 51.65-80, two groups are established for commercial condensates: I - for condensate stabilization units, II - for fields.

The main quality indicator of stable condensate is saturated vapor pressure, which characterizes the presence of light hydrocarbons in it. This indicator for Group I of products is (Table 3.1.1) for winter and summer periods of the year 93325 and 66661 Pa, respectively, for Group II-93325 Pa.

Norms for water and mechanical impurities content in condensate are set based on the requirements for normal storage and pumping of the product, as well as taking into account its further processing.

To fully assess the marketability of condensates, it is also necessary to determine such indicators as fractional composition, content of sulfur compounds, aromatic hydrocarbons and high-boiling paraffins, pour point, etc.

CHAPTER III

Based on the requirements to stable gas condensates, the developed technology of gas condensate stabilization meets the norms of this standard (OST 51.65-80).

CHAPTER IV.TECHNO-ECONOMIC RECOMMENDATIONS.

1. techno-economic recommendations on development of gas condensate stabilization process

The quality and stability of incoming oils and gas condensates are of great importance for satisfactory operation of refineries. They are often accompanied by very light hydrocarbons, such as propane-butane and C_5-C_6, which complicate the operation of the column of their primary processing (increase the partial pressure of the top of rectification columns, vapors by 10-15 atm), which is unacceptable and leading to severe consequences.

Therefore, the proposed work develops a moderate mode of the process of stabilization of gas condensate feedstock that meets the requirements of the refinery. It is proposed to reduce the temperature of vapor condensation of the upper part of the rectification column of gas condensate stabilization by 5-8° C and increase the phlegm number to 3^x light hydrocarbons coming to irrigation. In this case, irrigation should be performed not on the 35th plate, but on the 30th plate of the gas condensate stabilization column.

Table 19.

Calculation unit cost of stable gas condensate

Name	Unit.	Price of units, in UZS, thousand UZS	Consumption rate in t.	Stab. gas condensate price in sums	
				From USC-2.	Under the proposed
1	**2**	**3**	**4**	**5**	**6**
Unstable HC	A	7000	1,15	8050,0	8050,0
GC light (return)	A ton	926,0	0,15	-	-1368,9
Return n/a	A ton	7400	0,1	740	217,65
Fuel-Efficiency Costs: cooling water	Conditionally A ton	3	5+ 0,8 0,1	15 3880	24 2800

Continued in Table 19

Other expenses 10%		-	-	805	805
Overheads 15%		-	-	1207,5	1207,5
Total: factory cost of sales				14697,5	11735,25

As a result, with the new-introduction in the parameters of the technology of stabilization of gas condensate is achieved improvement of quality indicators of raw materials sent for processing and improves the operation of the propane-butane separation unit. Thus we will point out technical and economic indicators of a unit of stable gas condensate (table 19).

Economic efficiency per 1 ton of stable condensate by the proposed method turned out to be equal to 2962.25 sum.

Thus, at reduction of costs per ton of stable gas condensate, at the expense of propane-butane return from effective condensation of gas condensate in the amount of 217.65 sums its cost price is reduced by 2962.25 sums.

Recommended:

- development of calculation and design names in the condenser-dephlegmator part of the USK rectification column;

- working out of measures on the corner of change and creation of work instructions on the condenser part of the USC;

- development of approximate calculation indicators in real life on the operating plant.

CHAPTER IV

As can be seen from the calculations, our proposed technology of modernization of the USK unit is more effective in obtaining propane butane fraction. In addition, the application of this technology will reduce the plant production cost of propane butane fraction and stable gas condensate.

CONCLUSION

1. Based on the study of scientific, practical and production experience of a number of gas condensate stabilization units at GPP operated in natural gas fields it was revealed that the existing units do not achieve a clear final separation of light hydrocarbons (up to 8 % of propane-butane, up to 2-3 % of light hydrocarbon fraction).
2. At development on the laboratory stand of rectification process of separation of dissolved gases from gas condensate it is established that two or three times cooling (up to 8° -10° C) of upper fractions and increase of phlegm number (up to 3) for irrigation of the column top is achieved more clear separation of propane-butane from gas condensate.
3. The upper gas-extracting part of gas-condensate stabilization rectification column has been studied and calculations have been made.
4. With the introduction of the proposed parameters in the technology of rectification stabilization, the work of propane butane separation is unloaded and energy costs are reduced, which reduces the cost of stable gas condensate production by 5-6%.
5. The developed technology, more clearly regulating the parameters of the regime of the operating installations of gas condensate stabilization by introducing a newly introduced mode of in-line heating of incoming HC volumes and double use of light hydrocarbon phlegm for the purpose of more effective separation in the rectification column of associated gases and its qualitative stabilization, without significant changes in stabilization technology can be mastered in production practice at the GPP.

LIST OF REFERENCES USED.

1. Sayfullin et al. Automobile gasoline Euro-Super-95 AO "NOVIL" Petroleum Refining and Petrochemistry, 7 and 8, 2006. pp. 90-97 and 67-76.

2. Gureev.A.A. Application of automotive gasoline.// M.: Khimiya, 2012.364 p.

3. Mirsagatova M.A., Alimov A.A., Akhmedov K.S. Reaction - ability of gas condensate hydrocarbons to oxidation on oxide catalyst / Uzbek.chem.zh. 2000, №1. c. 25-28.3. Oil and gas industry of Uzbekistan - the basic branch of economy of the country // NHC "Uzbekneftegaz" T. - "Puls iks" 2004. C. 82.

4. Alimov A.A. // On chemical utilization of natural gas and gas condensates// Uzbek Chemical Journal. Tashkent, "Fan", 2003. №1. C. 87 - 94.

5. ABB Lummus Cross SHGHK Technological Regulations "Production of ethylene" polyethylene. // *Dosteinde AH voorburg The Netherlands USA* . 1997, 2272

6. Alimov A.A. Chemical processing of gas and gas condensate //Uzbek Chemical Journal 1993. №1. C. 45-52.

7. Samukov T. I., Khaltaev X. F., Alimov A. A. Obtaining individual solvents from gas condensate. // Uzbek Journal of Oil and Gas. 1997. №4 C. 48-50.

8. Abduganiev A. B., Alimov A. A. Obtaining conditioned toluene from gas condensate. // Uzbek Journal of Oil and Gas. 1999. № 3. C. 32-34.

9. Alimov A. A., Saitdinov F. A. Gas-condensate-crude for obtaining motor fuels // Uzbek Journal of Oil and Gas. 1999. № 4 C.30-31.

10.Timerkhanov F.Sh. The problem of soot reduction and improvement of technical and economic indicators of engines by using ashless additive TIOFAT in fuel. // Information leaflet no. 71-003-03. Kazan: Tatar CNTI.-2002.-4 pp.

11. Alimov A. A. Ismatov D. N. et al. Studies of kinetics of synthesis of iso-ethers of ashless additives of motor fuels. // Proceedings of the Republican Scientific and Technical Conference. "Actual problems of chemistry and chemical technology". Tashkent. 2002. C. 6-8.

12. Kontsova L.V., Chernousova N.N. High-boiling solvent of cellulose ethers. cellulose ethers. // Paint and varnish materials and their application. M.: 1981. № 5, - C. 59.

13. Karimov X. X. Machinosozlar uchn erituvchi olish. //Materials of scientific and practical conference of young scientists, teachers and professors. TashHTI-Tashkent 2002; P.110-113.

14. Ismatov D. I., Alimov A. A. et al. Development of technology for production of iso-ether additives of motor fuels. //Uzbek Journal of Oil and Gas. 2002. №2. C.22-24.

15. Raskatov V.M. Chuenkov V.S., Bessonova N.F., Weiss D.A. Machine-building materials: A brief reference book // M.: Mashinostroenie, 1980. C. 511

16. Samukov T. I. Obtaining and research of properties of solvents from gas condensate and development of their technologies. // Dissertation for the degree of Candidate of Technical Sciences. Tashkent, 1998. C.50-57.

17. Bekirov T.M. Primary processing of natural gases. M.: 1987, P.256

18. Luntovsky E.A., Salashnik M.M., Krasnikov A.A. Stabilization of gas condensate. Moscow: VNIIEGazprom, 1979. C.67

19.Gnusova S.P., Bergo B.G., Fishman L.L. Technical progress in the technology of gas condensate collection and stabilization. Moscow: VNIIEGazprom. 1977. C.57

20.Shulga V.A., Novikov P.P., Ovchenkov I.M. and others // Preparation and processing of gas and gas condensate. Moscow: VNIIEGazprom. 1980. № 10. C.28 - 29.

21.Tyulyandina K.A., Burlachenko G.M. Preparation and processing of gas and gas condensate. Moscow: VNIIEGazprom. 1982. № 6. C.10 - 12.

22. Bergo B.G., Frolov A.V., Fishman L.L., et al. // Improvement of gas condensate stabilization technology. Moscow: VNIIEGazprom. 1984. C.35

23.Bekirov T.M., Shkoryapkin A.I., Chernomyrdin V.N.// Peculiarities of development and operation of gas fields of the Caspian depression: Collection of scientific articles M.: VNIIEGazprom.1982.P.126-136.

24. Bekirov T.M., Stryuchkov V.M., Khalif A.L. et al. // Gaz. prom. M.: 1982 № 1. C. 33 - 34.

25.Bekirov T.M., Khalif A.L., Vezhnovets T.S., et al. // Preparation and processing of gas and gas condensate. Moscow: VNIIEGazprom. 1980. № 12. C.23 - 28.

26.Samukov T.I. "Obtaining and research of properties of solvents from gas condensate and development of their technologies" dissertation for the degree of candidate of technical sciences. T.: 1997, P.

27.Mirsagatova M.A. "Film formers based on oxidation products of gas condensate and their colloidal-chemical properties". // Abstract of Candidate of Chemical Sciences T.: 1987. C.23

28.Kunishev D. Regulation of colloidal and chemical properties of synthetic detergents with the help of hydrotronic substances derived from gas condensate. //T.: Abstract 1975, P.

27. Kozorezov Yu.N. "Processing of gas-condensate raw materials abroad: // Chemistry and technology of fuel and oils. 1966 №1 C.61-63

28. Bargo B.G., Gadzhiev N.B. Installation for investigation of condensate stabilization in a fractionating evaporator. // Gas and gas condensate processing. 1972. № 10 C.

29. Trivus N.A., Ovtina T.S. Experimental studies of partial stabilization of condensate at low-temperature separation units. // Gas and gas condensate processing. 1997. № 3. C.7-15.

31. Petrov N.A., Yuriev V.M., Hisaeva A.I. Synthesis of anionic and cationic surfactants for application in oil industry. Training manual UGNTU. - Ufa: 2008. C. 54

32. Bargo B.G., Gadzhiev N.B., Fishman L.L. Increase of productivity of light hydrocarbon fraction production. // Gas and gas condensate processing. 1976. № 11. C. 3-8.

33. Gnusova S.P., Baikova M.A. Stabilization of gas condensate in a rectification column. // Gas and gas condensate processing. 1975 № 11. C. 15-20.

34. Ruzmatov Sh.T., Turaev T.B. and Alimov A.A. Composition and properties of stable gas condensate after the stabilization unit // Proceedings of the International Conference "Catalytic Processes of Oil Refining, Petrochemistry and Ecology". Tashkent 2013. C. 287-288

35. Ruzmatov Sh.T., Muminov A.A., Turaev T.B. and Alimov A.A. Analysis of the process of stabilization of properties of gas condensate unitary subsidiary enterprise Shurtanneftegaz. // Bulletin Umidli kimyogar. Tashkent 2013g. C. 287-288

36. Technical regulations of gas condensate stabilization unit. UDP Mubarak

37. Technical Regulations of Gas Condensate Stabilization Unit USK-2 of UDP Shurtan

38. Alieva R.B. Processing and utilization of gas condensates. M.:1981. № 5.C.28 - 30.

39. Alimov A.A., Hashimov R., Hodzhakhanov N.A. Process technology for the preparation of alkyl - benzyltriethanolammonium chloride. //Collection of scientific works. T.: 1976. C. 46 - 55.

40. Kunishev D., Aminov S.N., Akhmedov K.S., Effect of hydrotropic substances on concentrated solutions of alkyl sulfate. // Journal of Colloid Chemistry. 1975 № 5 C.17-19.

41. Muratov T., Aminov S.N., Zainutdinov S., Akhmedov K.S. //Neftekhimiya i neftepererabotka. 1969. № 5. C. 54.

42. Makhmudov T.M., Alimov A.A., Ubaidullaev S., Akhmedov K.S. Synthesis of amines and cationic surface-active substances on the basis of gas condensate. //Synthesis and application of new surface-active substances. Tallinn 1973 P. 77 - 83.

43. Ivenyan A.I. Alimov A.A. Dehydration of gas condensate on sorbents. // Collection of scientific works. TashHTI-Tashkent, 2003: C.165-168.

44. Ivenyan A.I. Alimov A.A. Study of the process of purification of Kokdumalak gas condensate from dispersed mineralized water // Collection of scientific papers. TashHTI Tashkent, 2004, P.187-190.

45. Alimov A.A., Mirsagatova M.A. Oxidation of gasoline fraction of gas condensate. Uzbek chemical journal. VOL.: 1984 №5 P.39-42.

46. Alimov A.A., Mirsagatova M.A., Akhmedov K.S. Oxidation of gas condensate by air. T.: 1975, P.9.

47. Mirsagatova M.A., Alimov A.A., Akhmedov K.S. Reactivity of gas condensate hydrocarbons to oxidation on oxide catalyst. Uzbek Chemical Journal. T.: 1980. № 1. C. 25 - 28.

48. Alimov A.A., Dusmatov K.I. Oxidative ammonolysis of hydrocarbons of gas condensate.// Proceedings of Navoi Mining and Metallurgical Institute. Navoi., 2000, P. 211.

49. Mehdiyev S.D. Nitriles. // Baku Azerbaijan state edition. 1966. C. 467.

50. Alimov A.A. Chemical processing of gas condensates. Conference. "Actual problems of chemical processing of mineral raw material resources of Uzbekistan" T., 2003, P. 25.

51. Usheva N.V., Baramygina N.A. Research of gas condensate stabilization processes" .Tomsk- 2004: C. 3

52. Increase of efficiency of gas condensate field treatment technology. // Gas Industry.M.: 2003.No.7. - C. 54-57.

53. Mishin V.M. Processing of natural gas and condensate. // M.: Publishing Center "Academy", 1999. C. 448

54. http://www.ngpedia.ru/id474493p1.html. (Stabilization - Condensate).

55. http://nanoarea.ru/index.php/razlichnye-nauchnye-stati/258-tehnologija-stabilizatsii-kondensata-rektifikatsiej (Technology of condensate stabilization by rectification).

56. http://ntng.ru/ index1_7.html.(Gas condensate stabilization unit).

57. http://www.ngpedia.ru/ id548758p1.html. (Installation - stabilization - condensation)

58. http://www.himi.oglib.ru/bgl/80/120.htm l(Gas condensate processing technology)

59. http://vunivere.ru/work18118/page4 (Technique and technology of gas and condensate processing (Collection with the results of research of gas industry specialists obtained in the process of work)).

60. http://tekhnosfera.com /razrabotka-novyh-tehnologicheskih-resheniy-po-pererabotke-vysokoparafinistogo-

gazovogo-kondensata#ixzz32i9YUB7j (Development of new technological solutions for processing of highly paraffinic gas condensate).

61. http://www.findpatent.ru/patent/247/2477301.html (Method of processing unstable gas condensate and the installation for its implementation).

62. http://www.dissercat.com/content/otsenka-ekonomicheskoi-effektivnosti-ispolzovaniya-gazovogo-kondensata-v-rossii #ixzz32iBSmTrS (Assessing the economic efficiency of gas condensate utilization)

Printed by Books on Demand GmbH, Norderstedt / Germany